Power Systems Simulation

Power System Simulation

J.-P. Barret
Formerly Scientific Advisor
Power Systems Department
Research and Development Division
Electricité de France

P. Bornard
Manager Power Systems Department
Research and Development Division
Electricité de France

B. Meyer
Deputy Branch Manager Power System Economics and Design
Power Systems Department
Research and Development Division
Electricité de France

CHAPMAN & HALL
London · Weinheim · New York · Tokyo · Melbourne · Madras

Published by Chapman & Hall, 2–6 Boundary Row, London SE1 8HN, UK

Chapman & Hall, 2–6 Boundary Row, London SE1 8HN, UK

Chapman & Hall GmbH, Pappelallee 3, 69469 Weinheim, Germany

Chapman & Hall USA, 115 Fifth Avenue, New York, NY 10003, USA

Chapman & Hall Japan, ITP-Japan, Kyowa Building, 3F, 2-2-1 Hirakawacho, Chiyoda-ku, Tokyo 102, Japan

Chapman & Hall Australia, 102 Dodds Street, South Melbourne, Victoria 3205, Australia

Chapman & Hall India, R. Seshadri, 32 Second Main Road, CIT East, Madras 600 035, India

First edition 1997

© 1997 J.-P. Barret, P. Bornard and B. Meyer

Typeset in 10 on 12 pt Times by Best-set Typesetter Ltd., Hong Kong
Printed in Great Britain at T.J. Press (Padstow) Ltd., Padstow, Cornwall

ISBN 0 412 63870 3

Apart from any fair dealing for the purposes of research or private study, or criticism or review, as permitted under the UK Copyright Designs and Patents Act, 1988, this publication may not be reproduced, stored, or transmitted, in any form or by any means, without the prior permission in writing of the publishers, or in the case of reprographic reproduction only in accordance with the terms of the licences issued by the Copyright Licensing Agency in the UK, or in accordance with the terms of licences issued by the appropriate Reproduction Rights Organization outside the UK. Enquires concerning reproduction outside the terms stated here should be sent to the publishers at the London address printed on this page.

The publisher makes no representation, express or implied, with regard to the accuracy of the information contained in this book and cannot accept any legal responsibility or liability for any errors or omissions that may be made.

A catalogue record for this book is available from the British Library

Library of Congress Catalog Card Number: 96-86668

∞ Printed on permanent acid-free text paper, manufactured in accordance with ANSI/NISO Z39.48-1992 and ANSI/NISO Z39.48-1984 (Permanence of Paper).

Contents

Acknowledgements ix

1	**Introduction**	**1**
1.1	The study of electrical power systems	1
1.2	Simulation tools	2
	Further reading	5
2	**Modelling**	**6**
2.1	The role, importance and constraints of modelling	6
2.2	Models of knowledge and of behaviour	7
2.3	Data	11
	Further reading	11
3	**Steady state operation**	**12**
3.1	Introduction	12
3.2	System modelling	13
3.3	Power system equations	17
3.4	Load flow calculations	20
3.5	Direct current approximation	30
3.6	From load flow calculations to constrained optimization	33
	Appendix 3.A Recursive quadratic programming algorithms	46
	References	52
4	**Short-circuit currents**	**57**
4.1	Introduction	57
4.2	Definition of the short-circuit current	58
4.3	Method of calculating short-circuit currents	60
4.4	Modelling network elements for short-circuit calculations	66
	Further reading	78

5	**Long-term dynamics**	**79**
5.1	Introduction	79
5.2	Long-term dynamics model	80
	Appendix 5.A Representation of a classic thermal unit with a drum boiler	88
	Appendix 5.B Representation of a pressurized water reactor (PWR) nuclear unit	91
	Appendix 5.C Modelling a hydroelectric power unit	94
	Further reading	97

6	**Stability and electromechanical oscillations**	**98**
6.1	Introduction	98
6.2	Transient stability	99
6.3	Small signal stability	119
	Appendix 6.A Representation of the saturation	125
	Appendix 6.B Representation of faults	126
	Appendix 6.C The representation of rotating machines taking the dampers into account	129
	References	133
	Further reading	134

7	**Electromagnetic transients**	**137**
7.1	Introduction	137
7.2	Physical phenomena calling for modelling of electromagnetic transients	138
7.3	Types of modelling used	140
7.4	Method of solution	174
	Further reading	176

8	**Harmonics**	**177**
8.1	Introduction	177
8.2	Modelling of systems in harmonic conditions	178
8.3	Method of calculation	187
8.4	Propagation of harmonics in systems	188
8.5	Conclusion	202
	Appendix 8.A Representation of system components	202
	Appendix 8.B Distribution system	205
	References	206
	Further reading	207

9	**Digital real-time simulation**	**208**
9.1	Introduction	208
9.2	Real-time simulation for training dispatchers	209

9.3	Real-time simulation for tests on equipment	224
	Further reading	230

10 Computing facilities — **231**

10.1	Introduction	231
10.2	Information technology architecture	232
10.3	The graphical user interface	239
	Further reading	242

11 New developments — **243**

11.1	The coupling of different time-scales	243
11.2	Bringing together stability and long-term dynamics	247
11.3	New needs, new responses	248
	Appendix 11.A EUROSTAG integration methods	255
	Appendix 11.B Automatic differentiation method	257
	Appendix 11.C Application of a unique model of long-term dynamics and transient stability (EUROSTAG)	261
	Appendix 11.D A modelmaker-solver modelling application	263
	References	267
	Further reading	267

Appendix A: The modelling of direct current links — **270**

A.1	The composition of direct current links	270
A.2	Mode of operation and control of a station	270
A.3	Characteristics of a direct current station with $U_d = f(I_d)$	273
A.4	Mode of operation of two stations connected by a single direct current link	274
A.5	The modelling of direct current links	275
	Further reading	277

Index — 278

Acknowledgements

This book was made thanks to a collective effort and through many contributions from experts in the field of power system simulation. These colleagues, all belonging or having belonged to EDF Power Systems Department, have helped us through parts of this book, which would never have taken the present form without their work and dedication. The main contributors to this work are Jean-Claude Dodu (Chapter 3), Patrick Erhard (Chapter 7), Jacques Lachaume (Chapter 8), Michel Jerosolimski and Jérôme Macrez (Chapter 9) and Claude Bouquet (Chapter 10). Acknowledgements are also due to André Giard, Olivier Huet, Laurent Levacher, Michel Lemoine and Jean-Noël Marquet, among many others, for helpful contributions and proofreading.

Jean-Paul Barret, Pierre Bornard, Bruno Meyer
Clamart, September 1995

1

INTRODUCTION

1.1 THE STUDY OF ELECTRICAL POWER SYSTEMS

The design, construction and operation of a generating-transmission-distribution system for electrical energy are carried out with the triple aim of achieving quality of supply, safety of operation and economy.

To achieve this objective, it is necessary to understand thoroughly and to quantify, with the required accuracy, the phenomena which affect the electrical system. This is the role of power system studies, which cover a wide range of time constants.

This vision of time includes:

- steady state conditions in which all the variables and parameters are assumed to be constant during the period of observation: power distribution, sustained short-circuit current, etc.;

- slowly changing conditions corresponding to the normal changing patterns of loads and the action of automatic controls;

- transient conditions corresponding to the electromechanical oscillations of machines, and the action of primary voltage and speed controls;

- fast transient conditions in which the dimensions of the components of the power system are no longer negligible compared to the wavelength of the propagation phenomena.

Figure 11.1 shows the placing in time of phenomena relating to electric power systems.

This tracing of the slowest to the fastest phenomena through time will serve as the guiding thread throughout this work.

1.2 SIMULATION TOOLS

Since the beginning of electrical engineering in electric power systems, when investigating these different conditions, the engineer has encountered the difficulties of manual calculation through the sheer volume of operations required: complex networks, the division into a succession of steady states to examine dynamic phenomena, and so on.

To overcome this difficulty, various tools have been developed: reduced-scale models, analogue models, digital models and hybrid models. These different models are adapted to the problems dealt with in the context of certain hypotheses.

1.2.1 Reduced-scale models

These consist of elements representing the components of the power system at a certain scale. There will be generators, transformers, lines, underground cables, converters, loads, etc., connected as arranged on the actual system.

This category comprises:

- **Artificial systems**, representing a simple power system scheme sufficiently large to conduct tests on equipment (protection, automatic controllers, etc.). They simulate the electrical variables seen by these equipments under steady state and transient conditions.

- **Transient network analysers** for transient stability, representing complex networks intended to simulate dynamic stability phenomena. They include detailed models of generating units in which the motive element consists of a direct current motor, with the armature current controlled to reproduce the characteristics of the driving torque. The difficulty lies in constructing electrical machines which are 'homothetic' to real machines, in particular with regard to the time constants of the windings, the similarity being all the more difficult to achieve the lower the power of the model (a few kW). It may then be necessary to produce special machines and to resort to contrivances such as negative compensating resistances.

- **Transient network analysers for fast transients**, intended for simulating transient phenomena with wave propagation and frequencies up to a few kHz. They represent relatively small systems well, with a large number of elements, taking into account the frequencies involved.

- **Calculation tables**, consisting of resistors or impedances and power injection sources, can be considered as simplified reduced models of power systems. They were used formerly to calculate the distribution of power throughputs under steady state conditions: direct current tables consider only active power values, and alternating current tables are used for active and reactive power calculations. Today, they have completely disappeared, giving way to digital computers.

Reduced models have the advantage of operating in real time, and representing the physical reality of the phenomena. In this, they are excellent teaching tools. They are useful tools for research and the testing of equipment.

They have the following disadvantages:

- limitation of the size of the systems represented;
- lack of flexibility in changes of topology, although automatic connection systems have been created;
- considerable space is occupied;
- the need to produce specific equipment.

These various inconveniences mean that such tools are increasingly being replaced by digital computers.

1.2.2 Analogue models

In these models, electronic circuits such as operational amplifiers represent the behaviour of system components by solving their operating equations, for example Park's equations for alternators.

Today, although more versatile than reduced models, these calculation tools are being replaced by digital computers for the same reasons. However, they are still of interest for real-time simulation of high-speed phenomena which are too demanding for the current digital computers.

1.2.3 Digital models

The developments in calculating speed and memory size of digital computers have made it possible to produce efficient simulation tools with many advantages:

4 INTRODUCTION

- computers are standard equipment with standard systems;
- suitable calculation methods and algorithms have been created to reduce the calculation times, and to deal with large systems;
- data handling operations, in particular changes in hypotheses and topology, are facilitated and there are numerous possibilities for checking validity and consistency;
- the speed of calculation means that many situations can be considered, as in probabilistic planning studies;
- ease of processing the results for analysis;
- the possibility of working on a computer network to permit data sharing and access to powerful centralized or clustered calculation facilities.

Much continuing progress has been achieved in calculation time performances and storage capacity, releasing them from the constraints of data processing, and efforts are now concentrating on simulation techniques as such, opening up the range of phenomena which can be observed by one model and increasing the size of the systems studied.

New methods and algorithms well suited to digital calculation are under development.

A new approach, using modellers and solvers, allows algebraic-differential systems to be handled by bringing together function models.

1.2.4 Hybrid models

These models combine an analogue or reduced model and a digital computer to benefit from their respective advantages.

Their chief interest lies in real-time simulation, where the calculation of high-speed phenomena is too demanding for digital calculation and is entrusted to an analogue or reduced model. This is generally possible, since high-speed phenomena are concerned only with a limited part of the power system examined.

Another aspect of hybrid techniques lies in the use of digital computers to conduct the simulation and analyse the results of an analog or reduced model. These hybrid models, like the analogue ones, tend however to be replaced by digital ones.

1.2.5 Equivalent models

Whatever the simulation tool used, for research into very large systems we may be led to use equivalent systems of reduced size, intended to take the place of subsystems of a large system. The difficulty is then to ensure that the equivalent model does in fact have the same characteristics as the real power system with regard to the phenomena studied.

FURTHER READING

Anderson P.M. (1977). *Power System Control and Stability*, The Iowa State University Press.

Greenwood A. (1971). *Electrical Transients in Power Systems*, Wiley Interscience.

Sekine Y., Takahashi K. and Sakaguchi T. (1993). *Real-time Simulation of Power System Dynamics*, 11th Power System Computation Conference, Avignon.

2

MODELLING

2.1 THE ROLE, IMPORTANCE AND CONSTRAINTS OF MODELLING

Studies of electrical energy systems are based on the simulation of actual phenomena using models with the same behaviour as the elements in the physical system.

In practice, the electrical system has all the conditions presented in Chapter 1, sometimes simultaneously. An overall approach to these is obviously highly complex and today we do not know how to represent all the phenomena by one single form of modelling capable of giving a true image through the entire range of time constants.

Faced with a certain class of problems, the engineer needs a model suitable for investigating them. This need defines the characteristics of the model, and in particular the range of time constants to be covered.

With this in view, different models have been developed, each corresponding to a band which is necessarily limited with regard to time constant, subject to certain simplifications or approximations. All these constraints delimit an area of validity of the model in which it can be considered as truly representing the actual phenomena.

This is a fundamental point in simulation. One can only speak of a model by stating its field of validity defined by the hypotheses made for a given application. The fields of validity adopted cover the different conditions mentioned in Chapter 1.

One special aspect is real-time simulation, required for certain applications such as training simulators for operating personnel or simulators for tests on protection and automatic control equipment. This real-time aspect imposes an additional constraint on the speed of calculation of the phenomena simulated in order for the simulator to remain transparent for the operator.

2.2 MODELS OF KNOWLEDGE AND OF BEHAVIOUR

Depending on the type of research envisaged, a distinction can be made between two types of models leading to different performances.

2.2.1 Models of knowledge

For investigatory research, it is necessary to have models permitting precise and detailed simulation. The different parameters must be accessible.

These models follow the physical phenomena as closely and as faithfully as possible. They solve mathematical equations governing these phenomena. They are therefore generally complex and of considerable size. They constitute references in their field of validity.

2.2.2 Models of behaviour

These models must give precise results in a certain field of hypotheses corresponding to their use. These are 'black boxes' in which the concept of representation of the physical reality of the phenomena disappears, and only the relationship between data and results will count.

The models used for power system design and operation fall into this category. Their limited use leads to simpler models than the preceding ones, necessitating fewer data processing resources. This means that they can be more easily integrated into large simulation units, making it possible to cover a broad range of phenomena.

The case of modelling power station boilers is significant in this respect. For complete research into the physics of boilers and reactors, detailed models of knowledge are prepared. They demand on the one hand detailed input data, and on the other hand dedicated digital simulations. Electrical engineers have for a long time considered thermal boilers as a source of steam at constant pressure and temperature. This hypothesis is valid for typical phenomena in system design, concerning high-speed transients with regard to the time constants of these boilers. Initially, the models of boilers were sometimes non-existent.

Today, when simulating the electrical system, the engineer seeks to analyse the interaction between these boilers, turbines, alternators and the system as a whole. For research into some of these cases, models of behaviour of the black box type are sufficient. In fact, the full size of the system is often too large for it to be processed with detailed models of knowledge. Going beyond this, for more refined research which brings into play the interactions between the system and the internal protections of the boiler, a model of knowledge will have to be introduced to simulate the boilers.

We shall see (in Chapter 11) that techniques are emerging to bring in additional simulation tools. This makes it possible as a minimum to validate the models of behaviour in relation to the models of knowledge. In a further stage, it is possible to enhance the accuracy of the simulations by improving the quality of the modelling used.

Example: Model of 900 MW pressurized water reactors (PWR)

Lilliam is a 'simplified' dynamic behaviour model of the boiler of the PWR 900 MW French CP2 class of plant, developed for integration into power system simulation programs. A general representation of Lilliam is given in Figure 2.1.

The process representation is based on the fundamental physical laws of conservation and transfer of mass and energy. Though the model is simplified, its method of representation takes account of the principal nonlinearities inherent in the physical phenomena involved and of the main reactor limits and protections. The model is structured in functional modules (process, regulation, protections) to simplify subsequent upgrading and correction of the network.

The main aim of the model is to represent the principal dynamic variable of interface between the power plant and the network, namely the engine torque or its equivalent, the pressure at the steam barrel.

In order to further enhance the simulation of interactions between power plants and the network, the model integrates a detailed representation of certain internal protection and limiting systems of power plants liable to be activated by a network malfunction.

In addition, it provides a good representation of the upstream end of the turbine in large disturbances, making it possible to produce a faithful simulation of the significant major transients taken into account in power plant/grid interaction studies (particularly house load operation, separation from the network and behaviour after elimination of network malfunction).

To simulate correctly the activation of protections, it was necessary to develop extensions of the protections model to feature explicitly the physical quantities to which these protections react. Its main features are:

- representation of the characteristics of reactor coolant pumps and their driving engine and of the dynamics of flow rate build-up in the primary circuit;

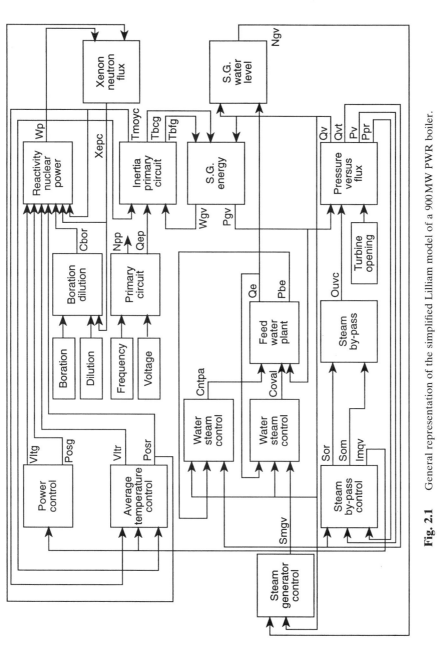

Fig. 2.1 General representation of the simplified Lilliam model of a 900 MW PWR boiler.

10 MODELLING

- complete representation of the feedwater plant (first, physical phenomena: mass balance at the steam generator including the contraction-expansion phenomenon, feedwater circuit with feedwater valves and feedwater turbopumps; and second the associated controls);
- representation (differentiation) of hot and cold sides of the primary circuit;
- integration of a two-zone neutron model to simulate the axial power imbalance over periods of a few hours.

The improvement of dynamic behaviour in large disturbances called for adapted and improved representation of certain sub-units of the model (in particular: adaptation of the steam bypass model; improvement of the reactor control rod model).

The response of the power plant (here the water level in the steam generator) when undergoing an islanding is represented in Figure 2.2. This simulation was performed using the Eurostag simulation package.

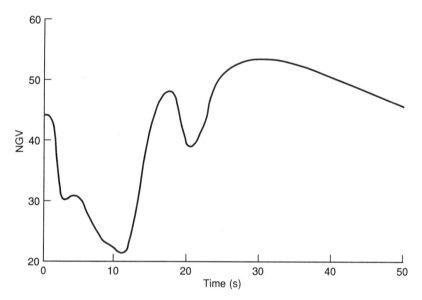

Fig. 2.2   Water level of the steam generator when the power station is undergoing an islanding.

2.3 DATA

'The best lifting tackle is only as strong as its weakest link' (Vauban). Similarly, the best model is valid only on account of the quality of the data supplied to it.

The problem of data, which is a key point in the simulation, has various aspects:

- Accessibility of the data – if knowledge of certain data does not present any difficulties, others are not easily accessible, either by calculation or by measurement, so it is appropriate to be reasonable when drawing up detailed models giving, in fact, illusory accuracy.

- The volume of data – it is interesting to examine the sensitivity of the results to data which can lead to simplification of the models without a marked loss of accuracy.

- Research into large systems, particularly under dynamic conditions, leads to the handling of a large number of data (systems with around ten thousand variables are now current). Apart from the problems of management and storage of these data, it is necessary to be certain of their validity and cohesion. The problem becomes more complicated when data arrive from different sources, for example from various electricity companies, not necessarily using the same definitions nor the same units. This leads to the problem of obtaining data from an external system which could be obtained by identification or by production of a model of equivalent size.

FURTHER READING

Anderson P.M. and Fouad A.A. (1977). *Power System Control and Stability*, The Iowa State University Press.

Chow J.H., Kokotovic P. and Thomas R.J. (1995). *Systems and Control Theory for Power Systems*, Springer-Verlag.

Del Toro V. (1992). *Electric Power Systems*, Prentice Hall.

Kundur P. (1994). *Power System Stability and Control*, McGraw Hill.

Meyer B., Lemoine M., Marquet J.N. *et al.* (1994). *Modelling of Power Plants for the Study of 'Plant/grid' Interactions under Dynamic Operating Conditions*, Cigré, 1994 Session, Paris.

Vernotte J.F., Panciatici P., Meyer B. *et al.* (1995). High fidelity simulation of power system dynamics. *IEEE Computer Applications in Power*, **8**, 1, January.

3

STEADY STATE OPERATION

3.1 INTRODUCTION

The general purpose of this chapter is the simulation of a power system in order to study its steady state operation. We need to know the steady state of a power system at a given moment, either to examine if the generation and transmission capacities are adapted to demand in a given situation, or to have a starting point (initial state) for transient state simulations such as the ones described in the following chapters.

The problem that we seek to solve is that of determining the steady state power flows through the different transmission system components, or more generally, that of calculating the voltage magnitudes and the voltage phase angles at all the system nodes. In this framework, we have adopted a simplified modelling through which the whole generation-transmission-consumption system can be rapidly simulated.

The first main simplification consists in considering only the electric variables with an angular frequency ω corresponding to the fundamental frequency which will vary only slightly from the nominal frequency (50 Hz or 60 Hz). Fast transient phenomena are thus not dealt with, and the time constants in the transmission lines (a few tens of milliseconds), in the transformers and in the generators (a few hundreds of milliseconds) are so low that they can be neglected. No component of the power system will therefore be modelled using a differential equation. Currents, voltages, powers, impedances are thus expressed as complex variables in polar coordinates (voltages and currents) or Cartesian coordinates (in the case of impedances and active and reactive powers).

The second simplification consists in limiting the study to equivalent single-phase circuits which correspond either to the balanced states of a three-phase network, or to the positive, negative or zero sequence components of the unbalanced states. We will however notice that this assumption implies the existence of a triangular symmetry for the three phases of the power system components. If this assumption could not be accepted, it would be necessary to use a detailed modelling.

3.2 SYSTEM MODELLING

Nodal topology and detailed topology

A state of a power system is first defined by its topology, i.e. on one hand by the list of the components in operation at the time studied and, on the other hand, by the connections between these components. There is a distinction between elementary (or detailed) topology and nodal topology.

In the detailed topology, a substation is represented with its switching equipment (circuit breakers, isolators), busbar sections, busbar connections, etc. It can be represented by a graph in which the nodes are the internal connections, and the branches the switching devices in the closed position. Each connected part of the graph of a substation is an electrical node. It is thus assumed that all the elements of a connected part have the same voltage, which means that the impedances of these elements can be neglected.

The nodal topology can be defined by a graph, the nodes of which are the electrical nodes and the branches of which are the transmission system components (lines, cables, transformers). In the load flow calculations, the system is represented in the nodal topology.

3.2.1 Line and cable modelling

A line or a cable connecting two nodes i and k is modelled by a π equivalent circuit with a series impedance

$$\overline{z}_{ik} = z_{ik} \exp(j\zeta_{ik}) = r_{ik} + jx_{ik}$$

and a shunt admittance on the side i

$$\overline{y}'_{ik} = g_{ik} + jh_{ik}$$

This π circuit is symmetrical (Figure 3.1), and moreover we have

$$g_{ik} = g_{ki} = 0, \quad h_{ik} = h_{ki} = \frac{C_{ik}\omega}{2}$$

where $C_{ik}\omega$ denotes the susceptance.

14 STEADY STATE OPERATION

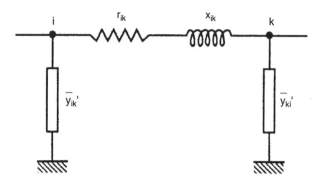

Fig. 3.1 Line and cable modelling.

3.2.2 Transformer modelling [1]

A transformer is classically represented by a diagram comprising a Γ circuit associated with an ideal transformer with a nominal voltage ratio $n_0 = n_1/n_2$, n_1 being the number of primary turns and n_2 the number of secondary turns (Figure 3.2(a)).

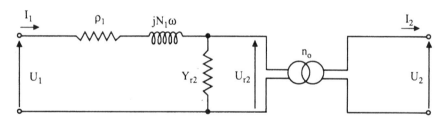

Fig. 3.2a Γ equivalent circuit of a transformer, referred to primary.

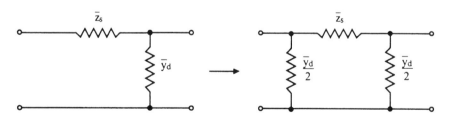

Fig. 3.2b Γ equivalent circuit and π equivalent circuit of a transformer.

In practice, the Γ circuit can be replaced by a π circuit which is symmetrical as for a line. If $\bar{z}_s$ and $\bar{y}_d$ denote the series impedance and the shunt admittance of the Γ circuit, respectively, we have (Figure 3.2(b))

$$\bar{z}_{ik} = \bar{z}_s, \quad \bar{y}'_{ik} = \frac{\bar{y}_d}{2}$$

This approximation is justified since z_d, the shunt impedence, is large in comparison with z_s.

Moreover, if the characteristics of the secondary network are referred to the primary, the ideal transformer need not be considered any longer. An impedance $\bar{z}_2$ of the secondary network is represented as 'referred to primary' by an impedance

$$\bar{z}_1 = \bar{z}_2 n_0^2$$

The secondary network voltages which are computed in this representation must be divided by n_0 to obtain the actual voltages.

Transformers with tap changing

The tap changer acts upon the primary winding

If n'_1 denotes the new number of primary turns, the transformer ratio becomes

$$n'_0 = t n_0, \quad \text{where } t = \frac{n'_1}{n_1}$$

The transformer can be represented by a diagram comprising a Γ circuit associated with two ideal transformers with ratios t and n_0, respectively (Figure 3.3). The characteristics $\bar{z}'_s$ and $\bar{y}'_d$ of the Γ circuit are no longer those, $\bar{z}_s$ and $\bar{y}_d$, of the transformer operating at the nominal ratio. When there is a modification of the transformer ratio, it is assumed that:

- the reactances vary in proportion to the square of the transformer ratio,
- the no-load admittances vary in proportion to the inverse of this ratio,
- the resistances vary in proportion to the square of the transformer ratio.

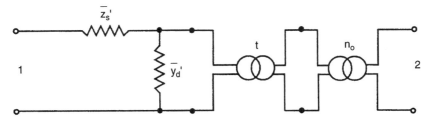

Fig. 3.3 Γ equivalent circuit of a transformer with tap changing when the tap changer acts upon the primary winding.

16 STEADY STATE OPERATION

The resistances vary in fact in a more complex manner according to the type of tap changer, but since their importance is very low in comparison with reactances, this approximation is justified. We therefore have

$$\bar{z}'_s = t^2 \bar{z}_s, \quad \bar{y}'_d = \frac{\bar{y}_d}{t^2}$$

Replacing the Γ circuit by a symmetrical π circuit and implying the ideal transformer with ratio n_o (which does not intervene in the calculations if the impedances are referred to primary), we obtain one of the three equivalent diagrams shown on Figure 3.4.

By comparing with Figure 3.2, we see in Figure 3.4(c) that the tap

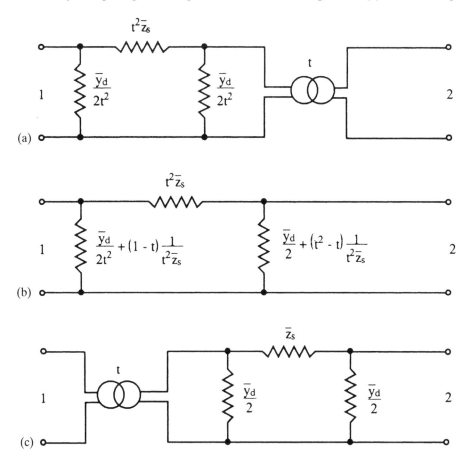

Fig. 3.4 Γ equivalent circuit of a transformer with tap changing when the tap changer acts upon the primary winding (alternative forms).

POWER SYSTEM EQUATIONS 17

changer only intervenes by adding an ideal transformer with ratio t on the primary side of the π equivalent circuit of the transformer studied, which has been established for nominal operation.

The tap changer acts upon the secondary winding

If n'_2 is the new number of secondary turns, then the transformer ratio becomes

$$n'_0 = \frac{n_0}{t}, \quad \text{where } t = \frac{n'_2}{n_2}$$

The transformer can be represented by the diagram shown on Figure 3.5. The tap changer only intervenes by adding an ideal transformer with ratio $1/t$ on the secondary side of the π equivalent circuit of the transformer studied, which has been established for nominal operation.

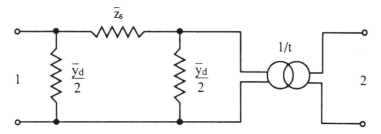

Fig. 3.5 Γ equivalent circuit of a transformer with tap changing when the tap changer acts upon the secondary winding.

3.2.3 Generation and consumption modelling

The generations and consumptions are represented by power injections into the electrical nodes. A positive or negative injection corresponds to a generation or a consumption, respectively. The algebraic sums of the active or reactive powers injected into the node i are denoted by P_i and Q_i, respectively.

3.3 POWER SYSTEM EQUATIONS

We will first examine the case in which transformers are not equipped with tap changers. We will then see the modifications caused by the presence of tap changers.

18   STEADY STATE OPERATION

3.3.1   Matrix equations

Let us consider the complex variables:

$\bar{I}_i$: injected current at node i ($i = 1, \ldots, n$)

$\bar{V}_i = V_i \exp(j\theta_i)$: voltage at node i

The current $\bar{I}_i$ can be written as a function of the voltages, thus

$$\bar{I}_i = \bar{y}_{ii} \bar{V}_i + \sum_{k \in \alpha(i)} \bar{y}_{ik}(\bar{V}_i - \bar{V}_k), \quad i = 1, \ldots, n$$

where $\alpha(i)$ denotes the subset of the nodes connected to node i and where

$$\bar{y}_{ii} = \sum_{k \in \alpha(i)} \bar{y}'_{ik}, \quad \bar{y}_{ik} = \frac{1}{\bar{z}_{ik}} = y_{ik} \exp(-j\zeta_{ik})$$

The preceding equations can be written in matrix form:

$$\bar{\mathbf{I}} = \overline{\mathbf{Y}} \overline{\mathbf{V}}$$

where $\overline{Y}_{ik} = G_{ik} + jH_{ik}, \quad \overline{Y}_{ii} = G_{ii} + jH_{ii}$

$$G_{ik} = -\frac{r_{ik}}{z_{ik}^2}, \quad H_{ik} = \frac{x_{ik}}{z_{ik}^2}$$

$$G_{ii} = \sum_{k \in \alpha(i)} (g_{ik} - G_{ik}), \quad H_{ii} = \sum_{k \in \alpha(i)} (h_{ik} - H_{ik})$$

3.3.2   Relations expressing the active and reactive powers as a function of the magnitudes and the voltage phase angles

Let P_i be the active power injected at node i, and let Q_i be the reactive power injected at node i. We have:

$$P_i - jQ_i = \bar{V}_i^* \bar{I}_i$$

by denoting by $\bar{V}_i^*$ the conjugate complex number of $\bar{V}_i$. Hence:

$$P_i = \mathrm{Re}\left\{ \bar{V}_i^* \left[\bar{y}_{ii} \bar{V}_i + \sum_{k \in \alpha(i)} \bar{y}_{ik}(\bar{V}_i - \bar{V}_k) \right] \right\}$$

$$Q_i = \mathrm{Im}\left\{\overline{V}_i^*\left[\overline{y}_{ii}\overline{V}_i + \sum_{k\in\alpha(i)}\overline{y}_{ik}(\overline{V}_i - \overline{V}_k)\right]\right\}$$

where Re and Im denote the real part and the imaginary part of a complex number, respectively. We can thus deduce:

$$\begin{cases} P_i = V_i^2 \sum_{k\in\alpha(i)}(y_{ik}\cos\zeta_{ik} + g_{ik}) - V_i \sum_{k\in\alpha(i)} V_k y_{ik}\cos(\zeta_{ik} + \theta_i - \theta_k) \\ Q_i = V_i^2 \sum_{k\in\alpha(i)}(y_{ik}\sin\zeta_{ik} - h_{ik}) - V_i \sum_{k\in\alpha(i)} V_k y_{ik}\sin(\zeta_{ik} + \theta_i - \theta_k) \end{cases} \quad (3.3.1)$$

These relations can be written in the form:

$$P_i = \sum_{k\in\alpha(i)} P_{ik}, \quad Q_i = \sum_{k\in\alpha(i)} Q_{ik}$$

where P_{ik} and Q_{ik}, respectively, denote the active and reactive power flows through the line connecting the nodes i and k. We have:

$$\begin{cases} P_{ik} = V_i^2(y_{ik}\cos\zeta_{ik} + g_{ik}) - V_i V_k y_{ik}\cos(\zeta_{ik} + \theta_i - \theta_k) \\ Q_{ik} = V_i^2(y_{ik}\sin\zeta_{ik} - h_{ik}) - V_i V_k y_{ik}\sin(\zeta_{ik} + \theta_i - \theta_k) \\ P_{ki} = V_k^2(y_{ik}\cos\zeta_{ik} + g_{ik}) - V_i V_k y_{ik}\cos(\zeta_{ik} + \theta_k - \theta_i) \\ Q_{ki} = V_k^2(y_{ik}\sin\zeta_{ik} - h_{ik}) - V_i V_k y_{ik}\sin(\zeta_{ik} + \theta_k - \theta_i) \end{cases} \quad (3.3.2)$$

3.3.3 Taking the tap changers into account

Let us study the case where the circuit connecting the nodes 1 and 2 is a transformer with tap changing. We will assume that the index 1 corresponds to the primary and the index 2 to the secondary, and we will denote by t the tap changer adjustment factor.

If the tap changer acts upon the primary winding (Figure 3.4(c)), V_1 just has to be replaced by V_1/t in the expressions of P_{12}, Q_{12}, P_{21}, Q_{21} as a function of the magnitudes and the voltage phase angles. If the tap changer acts upon the secondary winding (Figure 3.5), V_2 just has to be replaced by V_2/t in these same expressions.

In both cases, we can see that it is sufficient to divide by t the voltage magnitude of the node corresponding to the location of the tap changer, in the equations of the transformer operating at its nominal ratio.

3.4 LOAD FLOW CALCULATIONS

3.4.1 Definition

Let $N = \{1, 2, \ldots, n\}$ be the set of the indices of the power system nodes, and let U be a subset of N. We assume the voltage magnitude to be fixed for each node i belonging to U:

$$V_i = V_i^0, \quad \forall i \in U$$

In the absence of tap changers, the problem can be formulated as follows: knowing the injections of active power P_i ($\forall i \in N$) and the injections of reactive power Q_i ($\forall i \in N - U$), determine the values of θ_i ($\forall i \in N$) and of V_i ($\forall i \in N - U$). Taking into account the equations that were previously established, a solution to the problem is obtained by solving the nonlinear system:

$$P_i = G_{ii} V_i^2 - \sum_{k \in \alpha(i)} V_i V_k y_{ik} \cos(\zeta_{ik} + \theta_i - \theta_k), \quad \forall i \in N$$

$$Q_i = -H_{ii} V_i^2 - \sum_{k \in \alpha(i)} V_i V_k y_{ik} \sin(\zeta_{ik} + \theta_i - \theta_k), \quad \forall i \notin U$$

where $V_i = V_i^0$, if $i \in U$. This system has $2n - p$ equations and $2n - p$ unknowns, where p is the number of nodes belonging to U. Its solution makes it possible to calculate Q_i ($\forall i \in U$) by using the equations

$$Q_i = -H_{ii} V_i^2 - \sum_{k \in \alpha(i)} V_i V_k y_{ik} \sin(\zeta_{ik} + \theta_i - \theta_k)$$

The power flows through the transmission circuits, P_{ik}, Q_{ik}, P_{ki}, Q_{ki}, can also be deduced from it.

The nodes $i \in U$ are called generation nodes or PV nodes. At these nodes, P_i and V_i are known and we seek to determine Q_i and θ_i. We assume that the generation units connected to these nodes are equipped with voltage regulators which maintain the voltage magnitude V_i constant and equal to a rated value V_i^0. It is thus assumed that these units have reactive power margins that are sufficient in comparison with their supply or absorption limits. The nodes $i \in N - U$ are called consumption nodes or PQ nodes. At these nodes, P_i and Q_i are known and we seek to determine V_i and θ_i.

Taking the transformer tap changers into account can be done by considering as variables the tap changer adjustment factors and by taking as constant the controlled voltage magnitudes. Knowing the active injections at each node, the reactive injections at the nodes $i \in N - U$ and the voltage magnitudes at the nodes $i \in U$, our aim is to determine the voltage phase angles at each node, the voltage magnitudes at the nodes $i \in N - U$ where the voltage is not controlled, and the tap changer adjustment factors.

Let us consider for instance a transformer with tap changer connecting the nodes 1 and 2, the index 1 corresponding to the primary and the index 2 to the secondary. Let us assume that the tap changer acts upon the primary winding and let us denote by t the tap changer adjustment factor. In this case, the node 2 is the controlled node and V_2 is set. The unknowns of the problem which are associated to node 2 are θ_2 and t, instead of θ_2 and V_2. The total number of unknowns is thus unchanged, i.e. $2n - p$, if we assume that the tap changers of the transformers in parallel operate in the same manner.

With or without tap changers, the system to be solved is thus a system of $2n - p$ nonlinear equations with $2n - p$ unknowns. We will make below the two following simplifying assumptions:

- U is an empty set ($p = 0$).
- The system does not have transformers with tap changing.

The system to be solved can then be written in the form:

$$P_i = \varphi_i(\theta, V), \quad \forall i \in N$$
$$Q_i = \psi_i(\theta, V), \quad \forall i \in N$$

where θ is a vector of components $\theta_i (\forall i \in N)$ and V is a vector of components $V_i (\forall i \in N)$.

3.4.2 Properties of the nonlinear system [2]

If (θ, V) denotes a solution, then $(\theta + \sigma e, V)$, where σ is an arbitrary constant and e an n-dimensional vector with n components equal to 1, is also a solution. We can therefore select one of the nodes as the origin of the phase angles, i.e. set for instance $\theta_s = 0$. The node s is then called the reference node.

Let us consider the Jacobian matrix of the system (a $2n \times 2n$ square matrix):

$$\mathbf{J} = \begin{matrix} & (n) & (n) & \\ & \begin{bmatrix} \dfrac{\partial \varphi}{\partial \theta} & \dfrac{\partial \varphi}{\partial V} \\ \dfrac{\partial \psi}{\partial \theta} & \dfrac{\partial \psi}{\partial V} \end{bmatrix} & \begin{matrix} (n) \\ (n) \end{matrix} \end{matrix}$$

22 STEADY STATE OPERATION

J is a sparse matrix such that:

$$\frac{\partial \varphi_i}{\partial \theta_k} = \frac{\partial \varphi_i}{\partial V_k} = \frac{\partial \psi_i}{\partial \theta_k} = \frac{\partial \psi_i}{\partial V_k} = 0, \quad \text{if } k \notin \alpha(i), \ k \neq i$$

Moreover, the nonzero elements are located symmetrically with respect to the main diagonal (Remark 3.4.2.1).

The usual values of $\boldsymbol{\theta}$ and $\boldsymbol{V}$, corresponding to a stable state of the power system, are such that:

$$\sin(\theta_i - \theta_k) \neq \theta_i - \theta_k \ (k \in \alpha(i)), \quad V_i \simeq V_k \simeq V_{\text{nominal}} \quad (3.4.1)$$

Moreover, for a transmission system, the resistance r_{ik} of a circuit is small in comparison with the reactance x_{ik} (except for the cables) and the shunt admittance $\bar{y}'_{ik}$ is negligible. In these conditions and in the absence of cables, the nonzero elements of the submatrices $\partial \varphi / \partial V$ and $\partial \psi / \partial \theta$ are small in comparison with those of the submatrices $\partial \varphi / \partial \theta$ and $\partial \psi / \partial V$. If $\Delta \boldsymbol{P}$ and $\Delta \boldsymbol{Q}$ denote small variations of $\boldsymbol{P}$ and $\boldsymbol{Q}$ (n-dimensional vectors), we can then write:

$$\Delta \boldsymbol{P} = \left(\frac{\partial \boldsymbol{\varphi}}{\partial \boldsymbol{\theta}} \right) \Delta \boldsymbol{\theta}, \quad \Delta \boldsymbol{Q} = \left(\frac{\partial \boldsymbol{\psi}}{\partial \boldsymbol{V}} \right) \Delta \boldsymbol{V}$$

where $\Delta \boldsymbol{\theta}$ and $\Delta \boldsymbol{V}$ are the corresponding variations of $\boldsymbol{\theta}$ and $\boldsymbol{V}$ with respect to a solution to the nonlinear system.

The first n columns of **J** are not linearly independent. We have, for $i = 1, 2, \ldots, n$:

$$\sum_{k=1}^{n} \frac{\partial \varphi_i}{\partial \theta_k} = 0, \quad \sum_{k=1}^{n} \frac{\partial \psi_i}{\partial \theta_k} = 0$$

The matrix **J** is thus singular whatever the values of $\boldsymbol{\theta}$ and $\boldsymbol{V}$. Consequently, there exists a linear relation between the rows of **J**, which can be written

$$\mathbf{J}^{\text{T}} \begin{bmatrix} \boldsymbol{\lambda} \\ \boldsymbol{\mu} \end{bmatrix} = 0$$

where $\boldsymbol{\lambda}$ and $\boldsymbol{\mu}$ are two n-dimensional vectors (depending on $\boldsymbol{\theta}$ and $\boldsymbol{V}$) and where $\mathbf{J}^{\text{T}}$ is the transpose of **J**. For the usual values of $\boldsymbol{\theta}$ and $\boldsymbol{V}$, corresponding to a stable state of the system, **J** has rank $2n - 1$. **J** can have a lower rank when the system is in an unstable state, for instance when the transmissible power limit has been reached at a node. If **J** has rank $2n - 1$, $[\lambda^T,\mu^T]^T$ is the eigenvector of $\mathbf{J}^T$ associated with the eigenvalue **0**.

Let us consider a linear system of the form

$$\mathbf{J}\begin{bmatrix}\Delta\theta \\ \Delta V\end{bmatrix} = \begin{bmatrix}\Delta P \\ \Delta Q\end{bmatrix}$$

where $\Delta\theta$, ΔV are unknowns and ΔP, ΔQ are given. A necessary and sufficient condition for this system to have a solution is that ΔP and ΔQ satisfy the relation

$$\lambda^T \Delta P + \mu^T \Delta Q = 0$$

Remark 3.4.2.1

In the presence of tap changers, the Jacobian matrix of the system to be solved is no longer symmetric topologically and it has more zero elements. Let us consider a transformer with tap changing which connects the nodes 1 and 2, the index 1 corresponding to the primary and the index 2 to the secondary. The tap changer adjustment factor is denoted by t, and it is assumed that the tap changer acts upon the primary winding. We then have:

$$\frac{\partial \varphi_i}{\partial t} = \frac{\partial \psi_i}{\partial t} = 0, \quad \forall i \in \alpha(2) - \{1\}$$

although, in the absence of the tap changer:

$$\frac{\partial \varphi_i}{\partial V_2} \neq 0, \quad \frac{\partial \psi_i}{\partial V_2} \neq 0, \quad \forall i \in \alpha(2) - \{1\}$$

Properties of matrix $\frac{\partial \varphi}{\partial \theta}$

The sum of the n columns of matrix $\frac{\partial \varphi}{\partial \theta}$ being zero, this matrix is singular. Under the conditions (Remark 3.4.2.2):

$$0 < \zeta_{ik} + \theta_i - \theta_k < \pi, \quad \forall i \text{ and } \forall k, \; k \in \alpha(i) \tag{3.4.2}$$

it has the following properties:

$$\frac{\partial \varphi_i}{\partial \theta_i} > 0, \quad \frac{\partial \varphi_i}{\partial \theta_k} \leq 0$$

24 STEADY STATE OPERATION

Let us consider a submatrix of $\partial\varphi/\partial\theta$, deducted from $\partial\varphi/\partial\theta$ by deleting the row and the column with index s (s being chosen arbitrarily). Let us denote by A this $(n-1) \times (n-1)$ submatrix. Its elements are:

$$A_{ik} = \frac{\partial \varphi_i}{\partial \theta_k}, \quad \forall i \in N - \{s\} \text{ and } \forall k \in N - \{s\}$$

Since, for all i:

$$\frac{\partial \varphi_i}{\partial \theta_i} = -\sum_{k \in \alpha(i)} \frac{\partial \varphi_i}{\partial \theta_k}$$

we have, for all i:

$$A_{ii} \geq -\sum_{k \neq i} A_{ik}$$

A is thus a diagonally dominant matrix. It can be shown that A is an M-matrix the inverse of which is positive (all elements of A^{-1} are positive) (Remark 3.4.2.3 gives the definition of an M-matrix).

Remark 3.4.2.2

We assume here that we study transmission systems verifying the conditions (3.4.1) and for which the resistance of a circuit is small in comparison with the reactance. In these conditions, ζ_{ik} is close to $\pi/2$ and $|\theta_i - \theta_k|$ is lower than $\pi/4$. The conditions (3.4.2) are thus satisfied.

Moreover, the studied network is always assumed to be connected, that is, it cannot be split into two independent subnetworks.

Remark 3.4.2.3

A real square matrix A is an M-matrix if A is invertible, $A^{-1} \geq 0$ and $A_{ij} \leq 0$, for all $i, j, i \neq j$.

Properties of $\dfrac{\partial \psi}{\partial V}$

To simplify the notations, we have assumed that $p = 0$ (U is an empty set). Under this assumption, the matrix $\partial\psi/\partial V$ is ill-conditioned. Under the usual conditions (defined by Remark 3.4.2.2), $\partial\psi/\partial V$ is a nonsingular matrix with rank n. However, it is almost singular. Under the conditions (3.4.2), it has the following properties:

$$\frac{\partial \psi_i}{\partial V_i} > 0, \quad \frac{\partial \psi_i}{\partial V_k} \leq 0$$

but it is not diagonally dominant.

In practice, U is not the empty set. The Jacobian matrix of the system to be solved is then a matrix of the form

$$\mathbf{J} = \begin{bmatrix} \mathbf{J}^1 & \mathbf{J}^2 \\ \mathbf{J}^3 & \mathbf{J}^4 \end{bmatrix}$$

where $\mathbf{J}^1 = \dfrac{\partial \boldsymbol{\varphi}}{\partial \boldsymbol{\theta}}$ and $\mathbf{J}^4$ is the matrix with elements

$$\frac{\partial \psi_i}{\partial V_k}, \quad \forall i \in N - U \text{ and } \forall k \in N - U$$

If the set U is 'large enough', $\mathbf{J}^4$ is an M-matrix. The significance of the term 'large enough' depends on the system studied. For the 800 node French EHV system for instance, some tens of nodes with fixed voltages are sufficient in general to ensure a good conditioning of the matrix $\mathbf{J}^4$.

3.4.3 Solution of the nonlinear system

The method initially used was the Gauss–Seidel method. However, this method has poor convergence properties: it converges in a number of iterations which is proportional to the size of the power system. Presently, the Newton method or one of its variants is the method which has been universally adopted [2, 3].

At each iteration k, the power system equations are linearized around the current solution $(\boldsymbol{\theta}^k, \boldsymbol{V}^k)$. We are thus led to formulate the linear system:

$$\mathbf{J}^k \begin{bmatrix} \Delta \boldsymbol{\theta} \\ \Delta \boldsymbol{V} \end{bmatrix} = \begin{bmatrix} \Delta \boldsymbol{P} \\ \Delta \boldsymbol{Q} \end{bmatrix}$$

where $\Delta \boldsymbol{\theta} = \boldsymbol{\theta} - \boldsymbol{\theta}^k$, $\Delta \boldsymbol{V} = \boldsymbol{V} - \boldsymbol{V}^k$, $\Delta \boldsymbol{P} = \boldsymbol{P} - \boldsymbol{\varphi}(\boldsymbol{\theta}^k, \boldsymbol{V}^k)$, $\Delta \boldsymbol{Q} = \boldsymbol{Q} - \boldsymbol{\psi}(\boldsymbol{\theta}^k, \boldsymbol{V}^k)$, and where $\mathbf{J}^k$ is the value of the Jacobian matrix for $\boldsymbol{\theta} = \boldsymbol{\theta}^k$ and $\boldsymbol{V} = \boldsymbol{V}^k$. As we have just seen, such a system has a solution if and only if

26 STEADY STATE OPERATION

$$\left(\boldsymbol{\lambda}^k\right)^T \Delta \boldsymbol{P} + \left(\boldsymbol{\mu}^k\right)^T \Delta \boldsymbol{Q} = 0$$

where $[\boldsymbol{\lambda}^k, \boldsymbol{\mu}^k]^T$ is the eigenvector of $(\mathbf{J}^k)^T$ associated with the eigenvalue 0. $\boldsymbol{P}$ and $\boldsymbol{Q}$ being given, this relation cannot be satisfied. The linear system considered thus does not have a solution, which means that it is impossible to fix P_i ($\forall i \in N$) and Q_i ($\forall i \in N$).

The method commonly used to overcome this difficulty consists in deleting one of the equations $\varphi_i(\boldsymbol{\theta}, \boldsymbol{V}) = P_i$, then in solving the new system thus obtained with the help of the Newton method. The active power corresponding to the deleted equation will not be given any more, but it will be possible to calculate it *a posteriori* from the values found for $\boldsymbol{\theta}$ and $\boldsymbol{V}$. Moreover, the node corresponding to the deleted equation will be chosen as the reference node. If s is the index of this node, we will thus set $\theta_s = 0$. The new system to be solved can then be written in the form:

$$\begin{aligned} P_i &= \tilde{\varphi}_i\left(\tilde{\boldsymbol{\theta}}, \boldsymbol{V}\right), \quad \forall i \in N, \ i \neq s \\ Q_i &= \tilde{\psi}_i\left(\tilde{\boldsymbol{\theta}}, \boldsymbol{V}\right), \quad \forall i \in N \end{aligned} \quad (3.4.3)$$

where $\tilde{\boldsymbol{\theta}}$ is an $(n-1)$-dimensional vector having θ_i, $\forall i \in N$, $i \neq s$, as components. Assuming that this system has been solved, it becomes possible to calculate the active power injected at node s by means of the equation $P_s = \varphi_s(\boldsymbol{\theta}, \boldsymbol{V})$.

We must have:

$$\sum_{\substack{i=1 \\ i \neq s}}^{n-1} P_i + P_s = \pi$$

where π represents the active losses in the lines and transformers. The node s is called the slack node (or slack bus) because the value calculated for P_s thus ensures the balance between active generation and active consumption, this latter including losses.

At each iteration of the Newton method, the following linear system will have to be solved:

$$\tilde{\mathbf{J}}^k \begin{bmatrix} \Delta \tilde{\boldsymbol{\theta}} \\ \Delta \boldsymbol{V} \end{bmatrix} = \begin{bmatrix} \Delta \tilde{\boldsymbol{P}} \\ \Delta \boldsymbol{Q} \end{bmatrix}$$

where $\Delta \tilde{\boldsymbol{P}}$ denotes a vector of components $\Delta P_i (\forall i \in N, i \neq s)$ and where $\tilde{\mathbf{J}}^k$ is deduced from $\mathbf{J}^k$ by deleting the row corresponding to the equation $P_s = \varphi_s(\boldsymbol{\theta}, \boldsymbol{V})$ and the column corresponding to the variable θ_s. For the usual values of $\boldsymbol{\theta}$ and V, corresponding to stable states of the system, $\tilde{\mathbf{J}}^k$ is a nonsingular matrix with rank $2n - 1$. A standard solution method for the linear system consists in factorizing this matrix in the form of the product

$$\tilde{\mathbf{J}}^k = \mathbf{L}^k \mathbf{U}^k$$

where $\mathbf{L}^k$ is a lower triangular matrix and $\mathbf{U}^k$ is an upper triangular matrix. The system can then be solved very simply by successive eliminations [4, 5, 6].

The Newton method converges locally, that is, the starting point of the algorithm $(\boldsymbol{\theta}^0, V^0)$ must be close enough to the solution. Its interest lies in its convergence rate, which is quadratic. However, calculating and factorizing the Jacobian matrix at each iteration can be very consuming in computer time. Several techniques can be implemented to reduce this drawback:

1. The matrices $\partial\boldsymbol{\varphi}/\partial V$ and $\partial\boldsymbol{\psi}/\partial\boldsymbol{\theta}$ can be neglected, which leads to replacing the solution of a linear system of dimension $2n - 1$ by the solution of two linear systems of dimensions $n - 1$ and n, respectively. This approximation amounts to saying that the active powers are only slightly dependent on the voltage magnitudes and that the reactive powers are only slightly dependent on the voltage phase angles. This is all the more true since the phase angles remain small (lightly loaded systems) and since the ratios r_{ik}/x_{ik} are low (high voltage systems).
2. The Jacobian matrix can be calculated again only every ρ iterations, ρ being a given number (of the order of a few units).
3. It is possible to use a quasi-Newton method which consists in updating an approximation of the Jacobian matrix at each iteration, while preserving its sparsity (the so-called Sparse–Broyden method, for instance [7]). With a more elaborate Sparse–Broyden method, the two terms of the Jacobian matrix factorization can be updated directly, so that there is no need for a refactorization at each iteration [8].

These techniques can however deteriorate the convergence properties. The fixed Jacobian matrix method can converge linearly and even diverge. The convergence rate of the Sparse–Broyden method is no longer quadratic, although it remains superlinear.

Remark 3.4.3.1

The technique which consists in neglecting the matrices $\partial\boldsymbol{\varphi}/\partial V$ and $\partial\boldsymbol{\psi}/\partial\boldsymbol{\theta}$ in the Jacobian matrix calculation at each iteration of the Newton method is the standard so-called active-reactive decoupling technique [9]. Another decoupling technique can also be used to perform load flow calculations: the CRIC technique described in section 3.4.4.

Remark 3.4.3.2

In practice, U is not empty ($p \neq 0$). Under the usual conditions for transmission systems, the Jacobian matrix of the system to be solved is a nonsingular matrix of rank $2n - p - 1$. When the active-reactive decoupling technique is used to perform a load flow calculation, two linear systems of dimensions $n - 1$ and $n - p$, respectively, have to be solved at each iteration of the Newton method.

3.4.4 The CRIC decoupling technique

Another decoupling technique can be used to perform a load flow calculation: the CRIC technique (Calcul de Réseaux Implicitement Couplés) [10].

The standard active-reactive decoupling technique is based on the fact that, in the usual situations corresponding to a stable state of the power system, the variations in the active injections depend more on the phase angle variations than on the variations in the voltage magnitudes (which leads to neglect of $\partial \varphi / \partial V$ in the Jacobian matrix), and vice versa for the reactive injection variations (which leads to neglect of $\partial \psi / \partial \theta$).

The CRIC technique is based on another consideration: when the active power injections are fixed, the mean active power flows through the transmission circuits vary only slightly as a function of the voltage magnitudes. The calculation method consists first in expressing the injected reactive powers as a function of the voltage magnitudes and of the mean active power flows in the form

$$Q_i = q_i(\mathbf{V}, \mathbf{T}), \quad \forall i \in N$$

where $\mathbf{V}$ is the voltage magnitude vector with components V_i ($\forall i \in N$), and where $\mathbf{T}$ is the vector of mean active power flows, with components T_{ik} ($i \in N$, $k \in N$, $k \in \alpha(i)$). $\mathbf{T}$ has $2m$ components if m is the number of transmission circuits, and we have $T_{ik} = -T_{ki}$. T_{ik} is defined by

$$T_{ik} = \frac{1}{2}(P_{ik} - P_{ki}) \tag{3.4.4}$$

where P_{ik} and P_{ki} are given by:

$$P_{ik} = V_i^2\left(y_{ik}\cos\zeta_{ik} + g_{ik}\right) - V_iV_k y_{ik}\cos\left(\zeta_{ik} + \theta_i - \theta_k\right)$$
$$P_{ki} = V_k^2\left(y_{ik}\cos\zeta_{ik} + g_{ik}\right) - V_iV_k y_{ik}\cos\left(\zeta_{ik} + \theta_k - \theta_i\right) \quad (3.4.5)$$

It is shown below how to determine the functions q_i. The essential property of these functions is that the Jacobian matrix $\partial q/\partial V$ (of dimension $n \times n$) is a sparse matrix. As for the Jacobian matrix $\partial \psi/\partial V$, we have

$$\frac{\partial q_i}{\partial V_k} = 0, \quad \text{if } k \notin \alpha(i),\ k \neq i$$

Let us assume that the system to be solved has the form:

$$P_i = \tilde{\varphi}_i(\tilde{\boldsymbol{\theta}}, V), \quad \forall i \in N,\ i \neq s$$
$$Q_i = \tilde{\psi}_i(\tilde{\boldsymbol{\theta}}, V), \quad \forall i \in N$$

where $\tilde{\boldsymbol{\theta}}$ is the vector of components $\theta_i(\forall i \in N, i \neq s)$, i.e. the system (3.4.3).

The solution algorithm is the following: at the beginning of each iteration, a value of the voltage magnitude vector, V^0, is available. The following operations are performed:

1. By using the Newton method, the following system of nonlinear equations is solved:

$$P_i = \tilde{\varphi}_i(\tilde{\boldsymbol{\theta}}, V^0), \quad \forall i \in N,\ i \neq s$$

   where the unknown is $\tilde{\boldsymbol{\theta}}$ and P_i is given. We thus obtain a value of the voltage phase angle vector, $\boldsymbol{\theta}^0$, where $\theta_s^0 = 0$.

2. The mean active power flows are calculated by taking $\boldsymbol{\theta} = \boldsymbol{\theta}^0$ and $V = V^0$ in equations (3.4.4) and (3.4.5). Let T^0 be the obtained value.

3. By using the Newton method, the following system of nonlinear equations is solved:

$$Q_i = q_i(V, T^0), \quad \forall i \in N$$

   where the unknown is V and Q_i is given. Let V^1 be a solution to this system. If $\|V^1 - V^0\| < \varepsilon$ is small enough, the algorithm stops. If not, an additional iteration is performed taking for V^0 the value V^1.

As for the standard active-reactive decoupling technique, this technique only makes use of sparse $n \times n$ matrices.

Calculation principle of the functions q_i

The reactive equations can be written as

$$Q_i = \sum_{k \in \alpha(i)} Q_{ik}, \quad \forall i \in N$$

where Q_{ik} is given by

$$\begin{aligned} Q_{ik} &= V_i^2\left(y_{ik}\sin\zeta_{ik} - h_{ik}\right) - V_i V_k y_{ik}\sin(\zeta_{ik} + \theta_i - \theta_k) \\ &= V_i^2\left(y_{ik}\sin\zeta_{ik} - h_{ik}\right) - V_i V_k y_{ik}\sin\zeta_{ik}\cos(\theta_i - \theta_k) \\ &\quad - V_i V_k y_{ik}\cos\zeta_{ik}\sin(\theta_i - \theta_k) \end{aligned} \quad (3.4.6)$$

By definition, we have:

$$\begin{aligned} T_{ik} &= \frac{1}{2}\left(P_{ik} - P_{ki}\right) \\ &= \frac{1}{2}\left(y_{ik}\cos\zeta_{ik} + g_{ik}\right)\left(V_i^2 - V_k^2\right) + V_i V_k y_{ik}\sin\zeta_{ik}\sin(\theta_i - \theta_k) \end{aligned} \quad (3.4.7)$$

With the help of (3.4.7), we can express $\sin(\theta_i - \theta_k)$ as a function of V_i, V_k and T_{ik}. Then, $\cos(\theta_i - \theta_k)$ can be expressed as a function of V_i, V_k and T_{ik} by using the relation:

$$\cos(\theta_i - \theta_k) = \left[1 - \sin^2(\theta_i - \theta_k)\right]^{1/2}$$

It is then sufficient to replace $\sin(\theta_i - \theta_k)$ and $\cos(\theta_i - \theta_k)$ in (3.4.6) by the obtained expressions.

3.5 DIRECT CURRENT APPROXIMATION

In planning studies of transmission systems, it is usual to consider a great number of situations which are all different in respect of the demand level and of the availability of generation and transmission equipment. For this, several thousands of situations are selected by random sampling (Monte-Carlo method). In each situation, an optimal power flow calculation is performed in order to check if the generation and transmission equipment is adapted to demand (loss-of-load calculation) [11]. The optimal power flow has to respect network flow constraints under both normal and emergency operating conditions. Detecting the unsatisfied flow constraints over the network requires a great number of load flow calculations and, if the complete active-reactive network equations are used, the computing time becomes much too expensive. An elementary study of a minimum of 2000 situations would correspond to three hours of CPU time on a Cray YMP computer for an 840-node system.

It should be noted that, in practice, the precision brought about by the use of the complete active-reactive network equations is not justified for transmission system planning studies given the uncertainties which affect the data knowledge. In this context, a simplified model appears to be sufficiently judicious. The necessary computing simplification is based on the following considerations [12]:

The active power flows through the overhead line networks mostly depend on the voltage phase angles and not so much on the voltage magnitudes ($\partial \varphi / \partial \theta \gg \partial \varphi / \partial V$); in the absence of underground cable networks, we then assume that the voltage magnitude has the same value at each node: $V_i = V$, for all i.

Only the phase angles then vary, although with slight differences between neighbouring nodes. We thus consider that:

$$\sin(\theta_i - \theta_k) \cong \theta_i - \theta_k$$
$$\cos(\theta_i - \theta_k) \cong 1 - \frac{1}{2}(\theta_i - \theta_k)^2$$

Under these conditions, equations (3.3.1) can be written as:

$$P_i = V^2 \sum_{k \in \alpha(i)} g_{ik} + V^2 \sum_{k \in \alpha(i)} \left[x_{ik}(\theta_i - \theta_k) + \frac{1}{2} r_{ik}(\theta_i - \theta_k)^2 \right] \Big/ z_{ik}^2$$
$$Q_i = -V^2 \sum_{k \in \alpha(i)} h_{ik} - V^2 \sum_{k \in \alpha(i)} \left[r_{ik}(\theta_i - \theta_k) - \frac{1}{2} x_{ik}(\theta_i - \theta_k)^2 \right] \Big/ z_{ik}^2 \quad (3.5.1)$$

The ratio r_{ik}/x_{ik} is negligible for the transformers and of the order of an average of $\frac{1}{4}$ for the transmission lines. On the other hand, $(\theta_i - \theta_k)^2/2$ is very small in comparison with $(\theta_i - \theta_k)$, the phase angle difference between two neighbouring nodes being usually rather less than 10 degrees. Hence, in the expression of P_i, the term $r_{ik}(\theta_i - \theta_k)^2/2$ is negligible in comparison with the term $x_{ik}(\theta_i - \theta_k)$. Moreover, the terms g_{ik} are always very small (they corre-

32   STEADY STATE OPERATION

spond to the no-load active losses of the transformers). Finally, x_{ik}/z_{ik}^2 can be replaced by its approximate value $1/x_{ik}$. With these approximations, P_i can be simplified and becomes

$$P_i = \frac{V^2 \sum_{k \in \alpha(i)} (\theta_i - \theta_k)}{x_{ik}} \qquad (3.5.2)$$

In the same manner, the power flow P_{ik} through the branch (i,k) can be written as

$$P_{ik} = \frac{V^2 (\theta_i - \theta_k)}{x_{ik}} \qquad (3.5.3)$$

which is the expression of Ohm's law.

It should be mentioned that the assumptions made to obtain (3.5.2) and (3.5.3) correspond to a power system under normal operating conditions, excluding extreme phenomena such as voltage collapse or very high power flows, and for which the ratio r_{ik}/x_{ik} is small (this is the case for the high voltage systems). However, it is difficult to simplify Q_i in (3.5.1) under these assumptions.

Equations (3.5.2) and (3.5.3) can be written as:

$$\begin{cases} P_i = \sum_{k \in \alpha(i)} P_{ik} \\ P_{ik} = \dfrac{V^2 (\theta_i - \theta_k)}{x_{ik}} \end{cases} \qquad (3.5.4)$$

Equation (3.5.4) shows that, at each node i of the system, the generation or consumption (P_i) is balanced by what is imported or exported (P_{ik}). This power balance is the first Kirchhoff law if we consider the active power flow as analogous to a direct current.

Equation (3.5.3) is the expression of Ohm's law, where the phase angle difference is analogous to a potential difference in a direct current network (hence the name of the method) and, as previously, the active power flow is analogous to a current. The application of (3.5.2) along each independent mesh of the system gives the second Kirchhoff law or, in other words, the law of meshes.

The linear relation between **P** and **θ** makes it possible to solve the system of equations easily by using, for instance, the Zollenkopf method (bi-factorization) which takes advantage of sparsity [13]. In comparison with an active-reactive load flow calculation, a double saving of computing time is obtained: the problem to be solved is linear and the Newton method iterations are avoided. Moreover, the problem has a more reduced size (approximately half as much) since only active power flows are considered. Globally, the reduction in computing time is of the order of 100.

3.6 FROM LOAD FLOW CALCULATIONS TO CONSTRAINED OPTIMIZATION

For a great number of applications, the load flow calculation with fixed power injections only gives a partial answer to more general problems having an optimization requirement: for instance, to ensure a correct supply of the users and a good load dispatch by minimizing the operating costs of the generation units which each have a specific marginal cost which depends on the power output. Another problem consists in 'optimizing' the voltage profile in order to meet the equipment constraints, to avoid the risks for a voltage collapse, and to minimize the Joule losses or the amounts of reactive power compensation means to be installed.

In operation and planning studies of power systems, the optimization problems to be solved consist in minimizing a function of the variables P, Q, θ and V, while meeting the network load flow equations (3.3.1) and inequality constraints which express the operating limits of generation and transmision equipment (generation units, lines, transformers, etc.).

If 'security constraints' are not taken into account, such a problem can be written as a nonlinear mathematical program of the form:

$$\begin{cases} \text{Minimize } F(P_g, Q_g, \theta, V) \text{ subject to:} \\ \left. \begin{array}{l} P_{gi} - D_i = \varphi_i(\theta, V), \quad i = 1, 2, \ldots, n \\ Q_{gi} - E_i = \psi_i(\theta, V), \quad i = 1, 2, \ldots, n \end{array} \right\} \text{ i.e. equations } (3.3.1) \\ r_i(P_{gi}, Q_{gi}, V_i) \leq 0, \quad i = 1, 2, \ldots, n \quad (3.6.1) \\ h_j(\theta, V) \leq 0, \quad j = 1, 2, \ldots, m \quad (3.6.2) \end{cases}$$

where:
P_{gi} is the active power output of the generation units located at node i;
Q_{gi} is the reactive power supplied (or absorbed) by the generation units and the compensation devices (capacitors, reactors) located at node i;
D_i and E_i are the active and reactive consumptions at node i, respectively;
$r_i(P_{gi}, Q_{gi}, V_i) \leq 0$ represents local constraints such as voltage bounds, for

instance, or constraints which limit the operation of the generation units located at node i (bounds on the active power output, limit for absorption of reactive power, internal angle limit, stator current limit, etc.);

$h_j(\boldsymbol{\theta},V) \leq 0$, $j = 1, 2, \ldots, m$, represents the maximum transfer limits of the transmission circuits (lines, transformers).

The security constraints state that, in case of the loss of one or more generation units and/or transmission lines, the system will continue to operate within admissible limits. In section 3.6.3, we give a formulation of the voltage profile optimization problem, which takes into account these constraints.

The constraints (3.6.1) are most often bound constraints of the form

$$P_{gi}^m \leq P_{gi} \leq P_{gi}^M, \quad Q_{gi}^m \leq Q_{gi} \leq Q_{gi}^M, \quad V_i^m \leq V_i \leq V_i^M$$

where the superscripts m and M stand for minimum and maximum respectively.

The constraints (3.6.2) are of the form

$$I_{ik} \leq I_{ik}^M, \quad I_{ki} \leq I_{ki}^M \left(= I_{ik}^M\right)$$

where I_{ik} is the intensity, measured on the side i, of the current through the branch connecting the nodes i and k, and where I_{ki} is the intensity, measured on the side k, of the same current. I_{ik} and I_{ki} are such that

$$I_{ik}^2 = \frac{P_{ik}^2 + Q_{ik}^2}{V_i^2}, \quad I_{ki}^2 = \frac{P_{ki}^2 + Q_{ki}^2}{V_k^2}$$

where P_{ik}, Q_{ik} (respectively, P_{ki}, Q_{ki}) are the active and reactive powers through the branch connecting the nodes i and k, measured on the side i (respectively, measured on the side k) (Figure 3.6). Equations (3.3.2) give the expressions of P_{ik}, Q_{ik}, P_{ki} and Q_{ki}.

In 'optimal power flow' problems, the objective function to be minimized is the generation cost of thermal units. It has the form

$$F = \sum_{i=1}^{n} \Gamma_i\left(P_{gi}\right)$$

where the function Γ_i can generally be represented by a piecewise linear convex curve.

In 'voltage profile optimization' problems, the active power outputs of the generation units are set (except for active losses), and additional vari-

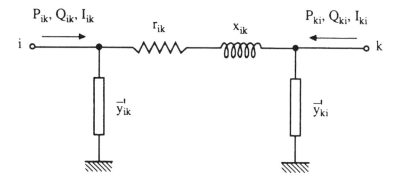

Fig. 3.6 Powers and currents through the branch (i,k).

ables $C_i \geq 0$ and $S_i \geq 0$ are introduced at each node i. These variables represent the additional amounts of capacitors and reactors, respectively, which will possibly be installed at node i to maintain the voltage within admissible limits. The network load flow equations are thus written as:

$$P_{gi}^0 - D_i + \lambda_i p = \varphi_i(\boldsymbol{\theta}, V), \quad i = 1, 2, \ldots, n$$
$$Q_{gi} - E_i + C_i - S_i = \psi_i(\boldsymbol{\theta}, V), \quad i = 1, 2, \ldots, n$$

where P_{gi}^0 and λ_i are given. The coefficients λ_i satisfy the relations

$$\lambda_i \geq 0 \; (i = 1, 2, \ldots, n), \quad \sum_{i=1}^n \lambda_i = 1$$

The coefficient λ_i represents the participation rate to the primary speed control of generation units located at node i. The variable p is such that

$$p = \pi - \sum_{i=1}^n (P_{gi}^0 - D_i)$$

where π represents the active losses in the transmission system. In fact

$$\pi^0 = \sum_{i=1}^n (P_{gi}^0 - D_i)$$

is an *a priori* estimation of active losses and p is the variation in active losses defined by $p = \pi - \pi^0$.

36 STEADY STATE OPERATION

The objective function to be minimized is generally of the form

$$F = \alpha p + \beta \sum_{i=1}^{n} C_i + \gamma \sum_{i=1}^{n} S_i$$

where α, β and γ are positive coefficients which represent the active loss cost, the investment cost for capacitors and the investment cost for reactors, respectively.

Instead of minimizing active losses, we can also seek to obtain the highest possible voltage profile. In this case, the objective function is of the form

$$F = \alpha' \sum_{i=1}^{n} \left(V_i^M - V_i\right) + \beta \sum_{i=1}^{n} C_i + \gamma \sum_{i=1}^{n} S_i$$

3.6.1 Solution methods

Since Carpentier's first formulation [14] in 1962 of the 'Optimal Power Flow' (OPF) problem, many solution methods have been used: first, penalty methods [15] and reduced gradient methods (particularly the generalized reduced gradient method [16]); then, more recently, recursive linear programming methods [17] and augmented Lagrangian methods [18]. Today, in a typical case such as the voltage profile optimization problem in planning studies, recursive quadratic programming methods are used. In these methods, the second-derivative matrix of the quadratic criterion is the Hessian matrix of the Lagrangian function of the problem (Newton-type methods [19, 20, 21, 22]) or an approximation of the Hessian matrix (quasi-Newton-type methods [23, 24]).

General principle

The recursive quadratic programming methods make it possible to solve nonlinear optimization problems of the form

$$\begin{cases} \text{Minimize } f(x) \text{ subject to:} \\ g_i(x) \leq 0, \quad i = 1, 2, \ldots, m \\ h_i(x) = 0, \quad i = 1, 2, \ldots, m \end{cases}$$

where x is a vector of $\mathbb{R}^n$ and f, g_i and h_i are functions from $\mathbb{R}^n$ to $\mathbb{R}$, which are assumed to be twice-continuously differentiable. In what follows, in order to simplify, we consider problems with only inequality constraints, of the form (P)

$$(P) \begin{cases} \text{Minimize } f(x) \text{ subject to:} \\ g_i(x) \leq 0, \quad i = 1, 2, \ldots, m \end{cases}$$

The iterative process of these methods is very simple. At the beginning of each iteration k, a point $x^k \in \mathbb{R}^n$ and an $n \times n$ symmetric matrix H^k are available. The following operations are performed:

1. A linearly constrained quadratic program, $Q(x^k, H^k)$, is solved. The constraints of this program are obtained by linearizing the constraints of the original problem around the current solution x^k. Its objective function is a second-order approximation of the original objective function. It is written:

$$\begin{cases} \text{Minimize } \nabla f(x^k)^T (z - x^k) + \frac{1}{2}(z - x^k)^T H^k (z - x^k) \text{ subject to:} \\ g_i(x^k) + \nabla g_i(x^k)^T (z - x^k) \leq 0, \quad i = 1, 2, \ldots, m \end{cases} \quad (3.6.3)$$

The solution to $Q(x^k, H^k)$ is denoted by z^k (we will see in Appendix 3.A that the choice of H^k ensures the uniqueness of this solution) and the vector of multipliers associated with the linear inequalities (3.6.3) at the optimum of $Q(x^k, H^k)$ is denoted by v^k.

2. x^{k+1}, the successor of the current point, is determined by performing a 'line search' on the line segment $[x^k, z^k]$. This line search consists in minimizing on $[x^k, z^k]$ a function which can be the function f or another (see the global convergence study in Appendix 3.A.1). The minimization can be exact or approximate. The line search can also consist in taking simply $x^{k+1} = z^k$.

3. H^{k+1} is determined as a function of H^k, x^k, x^{k+1} and v^k. According to which manner the matrix H^k and the line search are chosen, globally convergent algorithms or superlinearly and locally convergent algorithms are obtained (Appendix 3.A).

3.6.2 Voltage profile optimization and determination of reactive power compensation devices

The aim is to optimize the voltage profile of a power system and to determine the reactive power compensation devices (capacitors, reactors)

38 STEADY STATE OPERATION

which can prove necessary to maintain the voltage profile within admissible limits.

For this, a recursive quadratic programming method is used after writing the problem to be solved in the general form:

$$\begin{cases} \text{Minimize } f(x) \text{ subject to:} \\ g(x) = e \\ x^m \le x \le x^M \\ e^m \le e \le e^M \end{cases}$$

Except for the variable bounds, the constraints of the quadratic program solved at each iteration are linear equality constraints which result from linearizing the load flow equations about the current algorithm solution. It has the form:

$$Q(x^k, H^k) \begin{cases} \text{Minimize } \nabla f(x^k)^T (z - x^k) + \frac{1}{2}(z - x^k)^T H^k (z - x^k) \\ \text{subject to:} \\ g(x^k) + \nabla g(x^k)(z - x^k) = t \\ x^m \le z \le x^M, \quad e^m \le t \le e^M \end{cases} \quad (3.6.4)$$

by denoting by (x^k, e^k) the current algorithm solution (e^k does not intervene in the quadratic program) and by assuming that (x^k, e^k) verifies the bound constraints. The matrix H^k has the form

$$H^k = \nabla^2_{xx} \ell(x^k, v^k) + \alpha^k I$$

where $\ell(x, u)$ is the Lagrangian function

$$\ell(x, u) = f(x) + u^T g(x)$$

α^k is a nonnegative scalar and I is the $n \times n$ identity matrix (n being the dimension of x). Solving $Q(x^k, H^k)$ consists in determining a feasible solution (z^k, t^k) verifying the Kuhn–Tucker optimality conditions for $Q(x^k, H^k)$. v^k is the vector of multipliers associated with the constraints (3.6.4) at the optimum of $Q(x^k, H^k)$. The successor of (x^k, e^k) during the algorithm, i.e. (x^{k+1}, e^{k+1}), is determined by performing an Armijo-type line search on the line segment $[(z^k, t^k), (x^k, e^k)]$ (Appendix 3.A.1). The search function corresponding to this line search has the form

$$\theta_{r,u}(x, e) = f(x) + u^T [g(x) - e] + r \|g(x) - e\|_1$$

If r and u are chosen so that

$$r \geq \|v^k - u\|_\infty$$

the directional derivative of $\theta_{r,u}$ at (x,e) and in the direction

$$(d, \delta) = (z^k - x^k, t^k - e^k)$$

satisfies

$$\theta'_{r,u}(x,e;d,\delta) \leq -d^T H^k d$$

If $d^T H^k d > 0$, the line search can be performed. If not, the quadratic program $Q(x^k, H^k)$ has to be resolved, augmenting the value of α_k. The value of α_k becomes zero in a neighbourhood of a strict local minimum.

At each iteration, the values of r and u are updated with the help of an original technique due to Bonnans [25], which makes it possible to obtain a line search step length equal to 1 in a neighbourhood of a strict local minimum. The method thus obtained has the remarkable property of converging both globally and quadratically [26, 27]. Moreover, the implementation of the quadratic program is adapted to large-scale power systems. It is solved by using a reduced gradient method with LU factorization of the basis matrix and conjugacy of the reduced directions whenever this makes sense. The size of the studied systems can reach 2500 nodes.

3.6.3 Voltage profile optimization and determination of reactive power compensation devices taking into account security rules [28, 29]

The problem to be solved arises in planning studies of reactive power compensation devices. The aim is to optimize the amount (in MVAR) and the location of the capacitors and the reactors which are to be installed at each system node, so that operation security can be ensured under normal operating conditions as well as following contingencies which may disturb the normal operating conditions (loss of lines and/or generation units).

The problem can more precisely be defined in the following manner. A power system situation called the 'normal situation', as well as a list of contingencies which may affect this situation, are considered. The normal situation is characterized by a state of availability of generation and transmission equipment and by a consumption level. Each considered contingency results from the loss of one or more system components (lines, transformers, generation units). The problem consists in determining the global amounts, C_i and S_i, of the capacitors and reactors which are to be installed at each node i so that:

- there exists an admissible voltage profile (that is, a voltage profile which satisfies the voltage and reactive power limits) as well in the normal situation as in each of the considered contingency situations;
- the sum of the cost of active losses in the normal situation and of the investment cost for capacitors and reactors is minimal.

The two following assumptions are made:

- The active power outputs of generation units (except for active losses) are assumed to be set.
- It is assumed that the voltage profile after a contingency corresponds to the system operating state after the primary speed and voltage controls have acted, that is, a few seconds after the contingency.

In the case of a unit loss, the primary speed control is simulated assuming that the corrections to be made to the active power outputs under normal operating conditions vary linearly as a function of the active power output of the unit(s) which has (have) tripped. These corrections are problem data which are given by the planner.

The primary voltage control is modelled by imposing two conditions:

- The reactive power supplied (or absorbed) by the compensation devices is the same in the normal situation as in each contingency situation.
- The voltage at the unit terminals is the same in the normal situation as in each contingency situation, provided that the reactive power supply (or absorption) limits have not been reached in case of a contingency.

More precisely, let us denote by:

V_i the rated voltage of unit i (i.e. the voltage of unit i in the normal situation, which is an unknown of the problem);
V_i^k the voltage of unit i after contingency k;
Q_i^k the reactive power unit i after contingency k;
Q_i^m (respectively, Q_i^M) the limit for absorption (respectively, for supply) of reactive power.

We must have:

$$V_i^k = V_i, \quad \text{if } Q_i^m < Q_i^k < Q_i^M \tag{3.6.5}$$

$$V_i^k \leq V_i, \quad \text{if } Q_i^k = Q_i^M \tag{3.6.6}$$
$$V_i^k \geq V_i, \quad \text{if } Q_i^k = Q_i^m \tag{3.6.7}$$

Let us denote by G_k (respectively G_k^+, G_k^-) the subset of units verifying the relation (3.6.5) (respectively, (3.6.6), (3.6.7)). The subsets G_k, G_k^+, G_k^- (for all $k = 1, 2, \ldots, K$; K being the number of considered contingencies) are unknowns of the problem. Their determination poses a difficult combinatorial problem. In practice, the user makes a realistic estimation of the sets G_k, say G_k' ($k = 1, 2, \ldots, K$), and the following constraint is imposed:

$$V_i^k = V_i, \quad \forall i \in G_k' \text{ and } \forall k = 1, 2, \ldots, K$$

The method described below allows us to solve the problem posed, taking into account this simplifying assumption.

In this approach, the contingencies are taken into account in the preventive mode, that is, the control variables of the problem are determined so that, in case of a contingency and under the only action of automated mechanisms, the system continues to operate within admissible limits. The control variables (which are the components of the vector u defined below) are here the voltage magnitudes of the units belonging to the set

$$G = \bigcup_k G_k'$$

as well as the reactive power supplied (or absorbed) by the capacitors and the reactors. Other control variables can be introduced such as the adjustment factors of the transformer tap changers. In what follows, we will assume that all the sets G_k' are identical to G in order to simplify the writing.

The optimization problem to be solved can be written as:

$$(R) \begin{cases} \text{Minimize } F(x, u) \text{ subject to:} \\ \left. \begin{array}{l} p_i(x, u) = 0, \quad i = 1, 2, \ldots, n \\ q_i(x, u) = 0, \quad i = 1, 2, \ldots, n \end{array} \right\} & (3.6.8) \\ \left. \begin{array}{l} p_i^k(x^k, u) = 0, \quad i = 1, 2, \ldots, n \\ q_i^k(x^k, u) = 0, \quad i = 1, 2, \ldots, n \end{array} \right\} k = 1, 2, \ldots, K & (3.6.9) \\ \left. \begin{array}{l} u^m \leq u \leq u^M \\ x^m \leq x \leq x^M \end{array} \right\} & (3.6.10) \\ x^{m,k} \leq x^k \leq x^{M,k}, \quad k = 1, 2, \ldots, K & (3.6.11) \end{cases}$$

42 STEADY STATE OPERATION

where:

u is the vector of control variables (voltage magnitudes at the terminals of the units belonging to the set G, reactive power supplied (or absorbed) by the compensation devices);
x is the vector of the other variables in the normal situation;
***x**^{k}* is the vector of the other variables in the contingency situation k;
F is the objective function to be minimized (i.e. the sum of the cost of active losses in the normal situation and of the investment cost for compensation devices).

Equations (3.6.8) and (3.6.9) are the active-reactive load flow equations in the normal situation and in each of the considered contingency situations, respectively. Inequalities (3.6.10) and (3.6.11) represent the variable bounds.

When the control variables are fixed, the set of operating constraints in the contingency situation k, i.e. the constraints (3.6.9) and (3.6.11), has generally no feasible solution. For this reason, we add an artificial variable Δ_i^k at each node i and for each situation k. The variables Δ_i^k represent the reactive power supplied (or absorbed) by fictitious compensation devices. They must be zero at the optimum of the problem (R). In each situation k, we can thus define a function $f_k(\boldsymbol{u})$ as the optimal value of the objective function of the following optimization subproblem, where ***u*** is given:

$$P_k(\boldsymbol{u}) \begin{cases} \text{Minimize } \sum_{i=1}^{n} |\Delta_i^k| \text{ subject to:} \\ p_i^k(\boldsymbol{x}^k,\boldsymbol{u})=0, \qquad i=1, 2, \ldots, n \\ q_i^k(\boldsymbol{x}^k,\boldsymbol{u})+\Delta_i^k=0, \quad i=1, 2, \ldots, n \\ \boldsymbol{x}^{m,k} \leq \boldsymbol{x}^k \leq \boldsymbol{x}^{M,k} \end{cases}$$

$f_k(\boldsymbol{u})$ is an exterior penalty function with regard to the set of the vectors ***u*** which correspond to an admissible operating state of the power system in the contingency situation k. The function $f_k(\boldsymbol{u})$ is nonnegative and it is zero if and only if there exists an admissible operating state in the contingency situation k, when ***u*** is the value of the control variables. With the help of these functions, the optimization problem to be solved can be written in the form:

$$\begin{cases} \text{Minimize } F(\boldsymbol{x},\boldsymbol{u}) \text{ subject to:} \\ g(\boldsymbol{x},\boldsymbol{u}) \leq \boldsymbol{0} \\ f_k(\boldsymbol{u}) \leq 0, \quad k=1, 2, \ldots, K \end{cases}$$

CONSTRAINED OPTIMIZATION   43

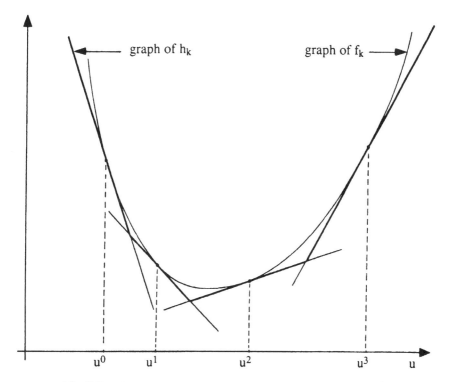

Fig. 3.7   Graph of h_k: a tangential approximation of f_k at iteration 4.

where $g(x,u) \leq 0$ represents the operating constraints in the normal situation, i.e. constraints (3.6.8) and (3.6.10).

The solution method proceeds by decomposition-coordination. As in [30, 31, 32], it is a Benders-type method in which a tangential approximation technique is applied to the penalty functions $f_k(u)$. At the iteration i of the solution algorithm, the function $f_k(u)$ is replaced by a tangential approximation (Figure 3.7):

$$h_k(u) = \max_{j \in J_i} \left\{ f_k(u^j) + (r_k^j)^T (u - u^j) \right\}$$

where $J_i = \{0, 1, \ldots, (i-1)\}$, u^j is the value of the control variables at iteration j, and where r_k^j is a subgradient of f_k at u^j, which is calculated from the Kuhn–Tucker multipliers at the optimum of $P_k(u^j)$. The current solution u^i is obtained by solving the following optimization subproblem (master program):

44 STEADY STATE OPERATION

$$\begin{cases} \text{Minimize } F(x,u) \text{ subject to:} \\ g(x,u) \leq 0 \\ h_k(u) \leq 0, \quad k = 1, 2, \ldots, K \end{cases}$$

or, in an equivalent form:

$$(\text{MP}_i) \begin{cases} \text{Minimize } F(x,u) \text{ subject to:} \\ g(x,u) \leq 0 \\ f_k(u^j) + (r_k^j)^T (u - u^j) \leq 0, \quad \forall j \in J_i \text{ and } \forall k = 1, 2, \ldots, K \end{cases} \quad (3.6.12)$$

The linear inequalities (3.6.12) are nothing but the cuts which are generated to ensure the feasibility of the solution when the generalized Benders method is applied [33].

The algorithm stops when, at iteration i:

$$\sum_{k=1}^{K} f_k(u^i) < \varepsilon \quad (3.6.13)$$

where ε is a given positive number which is chosen sufficiently small. If inequality (3.6.13) is satisfied, u^i is considered to be an optimal solution to (R). If not, iteration $(i+1)$ is performed (Figure 3.8).

The drawback of this method lies in the fact that the number of linear inequalities (3.6.12) increases by one unit per contingency at each iteration. The size of the master program (MP$_i$) can then become prohibitive. To overcome this difficulty, a relaxation technique can be used. At each iteration i, the linear inequalities (3.6.12) which have zero Kuhn–Tucker multipliers at the optimum of (MP$_i$) (that is, which are not binding) are eliminated from the master program.

The recursive quadratic programming method described in the previous section is the tool used to solve the master program (MP$_i$) as well as the optimization subproblems $P_k(u^i)$ at each iteration i. In the course of the solution of (MP$_i$), when the reduced gradient method is applied to one of the successive quadratic programs, the basis matrix and its inverse can be written as:

$$M = \begin{bmatrix} A & B \\ C & D \end{bmatrix}, \quad M^{-1} = \begin{bmatrix} A' & B' \\ C' & D' \end{bmatrix}$$

where A is a sparse nonsingular matrix and A' has the same dimension as A. The submatrices A and B correspond to a linearization of the load flow

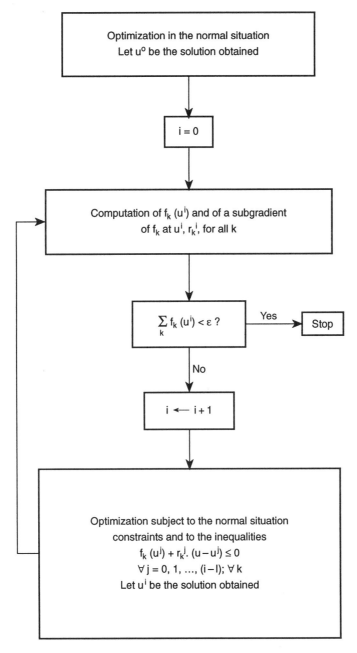

Fig. 3.8 Flow chart of the decomposition-coordination method.

equations, the Jacobian matrix of which is sparse. The submatrices C and D correspond to the linear inequalities (3.6.12), the Jacobian matrix of which is not sparse (all its elements are generally nonzero). The matrix A is factorized in the form of a product LU, L (respectively, U) being a lower (respectively, upper) triangular matrix, and the matrix D' is calculated with the help of the formula

$$D' = (D - CA^{-1}B)^{-1}$$

In the course of the iterations of the reduced gradient method, the factors L and U are updated by using the Reid technique [34, 35], and the matrix D' is updated by using the standard formulae of change of basis of the simplex method. It should be noted that the dimension of A, i.e. $2n \times 2n$, n being the number of system nodes, is, in practice, much higher than that of D'. In the numerical experiments which have been carried out on an 800-node EHV system with 20 contingencies taken into account, the number of inequalities (3.6.12) does not exceed 60 when the relaxation strategy is used, and the dimension of D' is thus at the most 60×60.

APPENDIX 3.A   RECURSIVE QUADRATIC PROGRAMMING ALGORITHMS

3.A.1   Globally convergent algorithms [36–41]

An algorithm converges globally if, whatever the starting point x^0, any limit point of the sequence of points $\{x^k\}_\mathbb{N}$ generated by the algorithm satisfies the Kuhn–Tucker necessary optimality conditions. In other words, any convergent subsequence of $\{x^k\}_\mathbb{N}$ converges to a point satisfying the Kuhn–Tucker optimality conditions.

To obtain this property, the matrix H^k is chosen (arbitrarily) in a compact set of positive definite symmetric matrices. The program $Q(x^k, H^k)$ is then a quadratic program with a strictly convex objective function. Assuming that the constraint set defined by the inequalities (3.6.3) is non empty, we deduce that its solution exists and is unique.

The function used for the line search is a function $\theta_r: \mathbb{R}^n \to \mathbb{R}$ defined by:

$$\theta_r(x) = f(x) + r\Phi[g(x)]$$

where $\Phi: \mathbb{R}^m \to \mathbb{R}$ is a convex penalty function and where r is a positive scalar. We take usually:

$$\Phi(y) = \sum_{i=1}^{m} \Phi_i(y_i), \quad \Phi_i(y_i) = \max\{0, y_i\}$$

In order to simplify, we use the notation x, H instead of x^k and H^k, and we denote by $\tilde{z}$ the solution to Q(x,H). Let $d = \tilde{z} - x$. The fundamental result is that the directional derivative of θ_r at x and in the direction d, is strictly negative if r is chosen large enough. More precisely, if $r \geq \|\tilde{v}\|_\infty$, we have:

$$\theta'_r(x;d) \leq -d^T H d$$

where the vector of multipliers associated with the inequalities (3.6.3) at the optimum of Q(x,H) is denoted by $\tilde{v}$. d is thus a descent direction of θ_r at x if $d \neq 0$ (if $d = 0$, x is a feasible Kuhn–Tucker solution for the original problem).

We can then proceed to a line search on the line segment $[x,\tilde{z}]$, θ_r being the descent function (Figure 3.A.1). It is a modification of the Armijo line

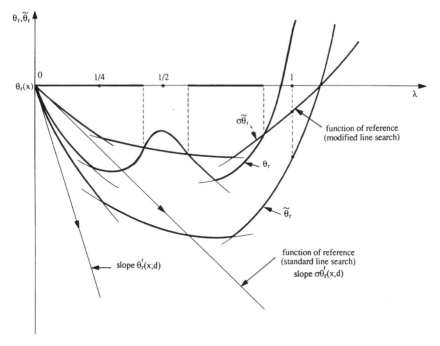

Fig. 3.A.1 Modified Armijo's line search where $\tilde{\lambda} = 1/4$. The values of λ satisfying equation (3.A.1) are shown by the thick lines.

search [42] where the linear function in λ, $\lambda\theta'_r(x;d)$, which is used as a function of reference, has been replaced by a convex function in λ, $\tilde{\theta}_r(x + \lambda d, x)$, defined by:

$$\tilde{\theta}_r(z,x) = f(x) + \nabla f(x)^T(z-x) + r\Phi\big[g(x) + \nabla g(x)(z-x)\big]$$

It can be verified that $\tilde{\theta}_r$ is indeed convex in z, x being fixed. Moreover:

$$\tilde{\theta}_r(x,x) = \theta_r(x)$$
$$\tilde{\theta}'_r(x,x;d) = \theta'_r(x;d)$$

The line search consists in solving the following problem:

$$\begin{cases} \text{Maximize } \lambda \text{ subject to:} \\ \lambda \in \Delta \\ \theta_r(x+\lambda d) - \theta_r(x) \leq \sigma\big[\tilde{\theta}_r(x+\lambda d, x) - \tilde{\theta}_r(x,x)\big] \end{cases} \quad (3.A.1)$$

where $\Delta = \{0\} \cup \{\lambda | \lambda = 2^{-i}, i \in \mathbb{N}\}$ and where $\sigma \in\ ]0,1[$. The solution to this problem, $\tilde{\lambda}(x,r,d)$, is determined by simply trying successively $\lambda = 1$, 2^{-1}, 2^{-2}, . . . This solution always exists and is unique. It is positive if $\theta'_r(x;d) < 0$. By setting $x' = x + \tilde{\lambda}d$, we have, if $r \geq \|\tilde{v}\|_\infty$:

$$\theta_r(x') - \theta_r(x) \leq -\sigma\tilde{\lambda}d^T H d (< 0 \text{ if } d \neq \mathbf{0})$$
$$x' \in F(x,r)$$

where F is a closed point-to-set map. These last two properties are essential for the global convergence of the algorithm which is defined as follows [43]:

Iteration k

x^k (feasible or not) and $r^{k-1} > 0$ are available:

- Choose H^k in a compact set of positive definite symmetric matrices, and solve $Q(x^k, H^k)$. Let z^k be the solution to $Q(x^k, H^k)$, let v^k be the vector of multipliers associated with the inequalities (3.6.3) at the optimum of $Q(x^k, H^k)$, and let $d^k = z^k - x^k$.
- If $z^k = x^k$, stop: x^k is a feasible Kuhn–Tucker solution.
- Otherwise, take:

$$r^k = \max\{r^{k-1}, \|\boldsymbol{v}^k\|_\infty\}$$

$$\boldsymbol{x}^{k+1} = \boldsymbol{x}^k + \lambda^k \boldsymbol{d}^k, \quad \text{where } \lambda^k = \tilde{\lambda}(\boldsymbol{x}^k, r^k, \boldsymbol{d}^k)$$

3.A.2 Superlinearly and locally convergent algorithms [44–51]

These algorithms are said to be locally convergent because the sequence of points generated by the algorithm is located in a sufficiently small neighbourhood of a strict local minimum $\hat{\boldsymbol{x}}$. It is generally assumed that $\hat{\boldsymbol{x}}$ is a regular point, that is, the gradients of the constraints which are active at $\hat{\boldsymbol{x}}$, $\nabla g_i(\hat{\boldsymbol{x}})$, for all $i \in \hat{E} = \{i \mid g_i(\hat{\boldsymbol{x}}) = 0\}$, are linearly independent. This assumption implies that there exists a unique vector of Kuhn–Tucker multipliers, $\hat{\boldsymbol{u}}$, associated with $\hat{\boldsymbol{x}}$. We will denote by E_+ the set

$$E_+ = \{i \in \hat{E} \mid \hat{u}_i > 0\}$$

by $\ell(\boldsymbol{x},\boldsymbol{u})$ the Lagrangian function

$$\ell(\boldsymbol{x},\boldsymbol{u}) = f(\boldsymbol{x}) + \boldsymbol{u}^\mathrm{T} \boldsymbol{g}(\boldsymbol{x})$$

and by $\hat{\boldsymbol{H}}$ the Hessian matrix of $\ell(\boldsymbol{x},\boldsymbol{u})$ at $(\hat{\boldsymbol{x}},\hat{\boldsymbol{u}})$

$$\hat{\boldsymbol{H}} = \nabla_{xx}^2 \ell(\hat{\boldsymbol{x}}, \hat{\boldsymbol{u}})$$

We assume that $\hat{\boldsymbol{H}}$ is positive definite along the subspace defined by

$$\hat{V} = \left\{ \boldsymbol{y} \in \mathbb{R}^n \,\middle|\, \nabla g_i(\hat{\boldsymbol{x}})^\mathrm{T} \boldsymbol{y} = 0, \ \forall i \in E_+ \right\}$$

that is

$$\boldsymbol{y}^\mathrm{T} \hat{\boldsymbol{H}} \boldsymbol{y} > 0, \quad \forall \boldsymbol{y} \in \hat{V} \ (\boldsymbol{y} \neq \boldsymbol{0})$$

This assumption implies that $\hat{\boldsymbol{x}}$ is a strict local minimum.

If we know a subset E of constraints which are active at $\hat{\boldsymbol{x}}$, such that $E_+ \subset E \subset \hat{E}$, we can replace the inequality constrained problem to be solved, (P), by the equivalent problem (P_E), with equality constraints

$$(P_E) \begin{cases} \text{Minimize } f(\boldsymbol{x}) \text{ subject to:} \\ g_i(\boldsymbol{x}) = 0, \quad \forall i \in E \end{cases}$$

$\hat{x}$ is also a strict local minimum of (P_E) and $\hat{u}_E$ is the vector of Lagrange multipliers which is associated with $\hat{x}$ at the optimum of (P_E).

If we know any point x that is close enough to $\hat{x}$, we can determine such a set E by simply solving a strictly convex quadratic program, $Q(x,H)$, with linear constraints

$$Q(x,H) \begin{cases} \text{Minimize } \nabla f(x)^T (z-x) + \frac{1}{2}(z-x)^T H(z-x) \text{ subject to:} \\ g(x) + \nabla g(x)(z-x) \leq 0 \end{cases}$$

where H is any positive definite symmetric matrix. E is then the index set of active constraints which have strictly positive multipliers at the optimum of $Q(x,H)$. The locally convergent iterative method which is considered to solve the original problem (P) practically consists in solving (P_E), after having thus determined E. At each iteration k, we will thus solve a quadratic program, $Q_E(x^k,H^k)$, with linear equality constraints

$$Q_E(x^k, H^k) \begin{cases} \text{Minimize } \nabla f(x^k)^T (z-x^k) + \frac{1}{2}(z-x^k)^T H^k (z-x^k) \\ \text{subject to:} \\ g_i(x^k) + \nabla g_i(x^k)^T (z-x^k) = 0, \quad \forall i \in E \end{cases}$$

In this quadratic program, the symmetric matrix H^k must be chosen, in a certain sense, as being close enough to $\hat{H}$. Let us denote by $V(x)$ the subspace defined by

$$V(x) = \left\{ y \in \mathbb{R}^n \middle| \nabla g_i(x)^T y = 0, \quad \forall i \in E \right\}$$

and let $P(x)$ be the $n \times n$ projection matrix on $V(x)$

$$P(x) = I - \nabla g_E(x)^T \left[\nabla g_E(x) \nabla g_E(x)^T \right]^{-1} \nabla g_E(x)$$

The pair (x^k, H^k) has to be such that

$$\left. \begin{array}{l} \|x^k - \hat{x}\| \leq \eta_0 \\ \|P(x^k)(H^k - \hat{H})\| \leq \eta_0' \end{array} \right\} \tag{3.A.2}$$

where η_0 and η_0' are two positive numbers which are small enough to have the following two properties:

- The matrix $\nabla g_E(x^k)$ has maximum rank.
- $y^T H^k y > 0$, $\forall y \in V(x^k)$ $(y \neq 0)$.

In these conditions, the matrix

$$\begin{bmatrix} H^k & \nabla g_E(x^k)^T \\ \nabla g_E(x^k) & 0 \end{bmatrix}$$

is nonsingular, which implies that the solution to $Q_E(x^k, H^k)$ is unique. It is the same for the Lagrange multiplier associated with this solution.

The algorithm is the following:

Iteration k

A pair (x^k, H^k) verifying (3.A.2) with H^k being symmetric is available.

- Solve $Q_E(x^k, H^k)$. Let z^k be the solution to $Q_E(x^k, H^k)$ and let v_E^k be the Lagrange multiplier associated with this solution.
- If $z^k = x^k$, stop: $x^k = \hat{x}$.
- If not, set $x^{k+1} = z^k$, and determine H^{k+1} with the help of an update formula of the form

$$H^{k+1} = H^k + \Phi\left(H^k, x^k, x^{k+1}, v_E^k\right)$$

Depending on the nature of H^k, we can have a Newton-type or a quasi-Newton-type algorithm. When $H^{k+1} = \nabla^2_{xx}\ell(x^{k+1}, u^k)$, where $u^k = [v_E^k, 0]$ (u^k being a vector of $\mathbb{R}^m$), the algorithm is said to be a Newton-type algorithm. When H^{k+1} is calculated by means of an update formula (such as Broyden, DFP, BFGS,...), the algorithm is said to be a quasi-Newton-type algorithm. Under the previous assumptions, a superlinear convergence rate is obtained by using the so-called 'double update' formulae [52, 53]:

$$H^{k+1} = P(x^{k+1}) H_1^{k+1} P(x^{k+1}) + P(x^{k+1}) H_2^{k+1} Q(x^{k+1}) + Q(x^{k+1}) H_2^{k+1} P(x^{k+1})$$

where

$$H_1^{k+1} = H^k + \Delta_1\left(H^k, a_1^k, b_1^k\right), \quad H_2^{k+1} = H^k + \Delta_2\left(H^k, a_2^k, b_2^k\right)$$

where $a_1^k = P(x^k)(x^{k+1} - x^k)$, $a_2^k = Q(x^k)(x^{k+1} - x^k)$
$b_1^k = \nabla_x \ell(x^k + a_1^k, u^k) - \nabla_x \ell(x^k, u^k)$
$b_2^k = \nabla_x \ell(x^k + a_2^k, u^k) - \nabla_x \ell(x^k, u^k)$

In this formula, $P(x)$ (respectively $Q(x)$) is the projection matrix on $V(x)$ (respectively, on $W(x)$ which is the orthogonal complement of $V(x)$). Δ_1 and Δ_2 correspond to two update formulae (m_1) and (m_2) of the form $\bar{H} = H + \Delta(H,a,b)$. If we moreover assume that $\nabla^2 f$ and $\nabla^2 g_i$ ($i = 1, 2, \ldots, m$) are Lipschitz continuous on a neighbourhood of $\hat{x}$, we have the following result: if DFP (respectively, BFGS) is chosen for (m_1) and Broyden for (m_2), then there exist four positive constants η, η', η_1, η_1' such that:

$$\left. \begin{array}{l} \|x^0 - \hat{x}\| \leq \eta_1 \\ \|P(x^0)(H^0 - \hat{H})\| \leq \eta_1' \end{array} \right\} \Rightarrow x^k \to \hat{x} \quad \text{if } k \to \infty, \; k \in \mathbb{N}$$

where $\|x^k - \hat{x}\| \leq \eta$, $\|P(x^k)(x^k - \hat{x})\| \leq \eta'$, $\forall k \in \mathbb{N}$

$$\frac{\|x^{k+1} - \hat{x}\|}{\|x^k - \hat{x}\|} \to 0 \quad \text{if } k \to \infty, \; k \in \mathbb{N}$$

Let us recall the Broyden, DFP, BFGS formulae [54–56]:

$$\bar{H}_1 = H + \frac{(b - Ha)b^T + b(b - Ha)^T}{a^T b} - \frac{a^T(b - Ha)}{a^T b} \frac{bb^T}{a^T b} \quad \text{(DFP)}$$

$$\bar{H}_1 = H + \frac{bb^T}{a^T b} - \frac{Haa^T H}{a^T Ha} \quad \text{(BFGS)}$$

$$\bar{H}_2 = H + \frac{(b - Ha)a^T}{\|a\|^2} \quad \text{(Broyden)}$$

REFERENCES

1. Villard P. (1974). Introduction des régleurs en charge des transformateurs dans les calculs de répartition et d'estimation de l'état du réseau. *Rapport EDF-DER*, HR 10671/3, March.

2. Feingold D. and Spohn D. (1967). Les problèmes de répartition de puissance dans les réseaux maillés à courant alternatif. Analyse et méthodes numériques. *Revue Générale de l'Electricité*, **76**(4), 681–96, April.

3. Tinney W.F. and Hart C.E. (1967). Power flow solution by Newton's method. *IEEE*, **PAS-88**, 1449–60.

4. Broussolle F. (1973). Calculs de répartition dans les réseaux électriques par la méthode de factorisation. *Rapport EDF-DER*, HR 10330/3, March.

5. Reid J.K. (1971). *Large Sparse Sets of Linear Equations*, Academic Press.

6. Sato N. and Tinney W.F. (1963). Techniques for exploiting the sparsity of the network admittance matrix. *IEEE*, **PAS-82**, 944–50.

7. Broyden C.G. (1971). The convergence of an algorithm for solving sparse nonlinear systems. *Mathematics of Computation*, **25**(114), 285–94.

8. Dennis J.E. and Marwill E.S. (1982). Direct secant updates of matrix factorization. *Mathematics of Computation*, **38**(158), 459–74.

9. Alsac O. and Stott B. (1975). Decoupled algorithms in optimal load flow calculations. *IEEE PES Summer Meeting*, Paper A75545-4, San Francisco.

10. Carpentier J. (1986). CRIC: A new active-reactive decoupling process in load flows, optimal power flows and system control. *Proc. IFAC Conference on Power Systems and Power Plant Control*, Beijing, August, 65–70.

11. Dodu J.C. and Merlin A. (1979). Recent improvements of the Mexico model for probabilistic planning studies. *Int. J. of Electrical Power and Energy Systems*, **1**(1), 46–56.

12. Augès P. (1963). Résolution des problèmes de répartition des puissances sur un réseau maillé. *Rapport EDF-DER*, HR 5143, April.

13. Zollenkopf K. (1971). Bi-factorization. Basic computational algorithm and programming techniques, in *Large Sparse Sets of Linear Equations*, Academic Press, London, 75–96.

14. Carpentier J. (1962). Contribution à l'étude du dispatching économique. *Bulletin de la Société Française des Electriciens*, **8**(3).

15. Sasson A. (1969). Nonlinear programming solutions for load-flow minimum-loss and economic dispatch problem. *IEEE*, **PAS-88**(4), 399–409.

16. Carpentier J. (1973). Differential injections method: A general method for secure and optimal load flow. *Proc. IEEE PICA Conference*, Minneapolis, 255–62.

17. Blanchon G., Dodu J.C. and Merlin A. (1983). Developing a new tool for real-time control in order to coordinate the regulation of reactive power and the voltage schedule in large-scale EHV power systems. *International Symposium on Control Applications for Power Systems*, Florence.

18. Burchett R.C., Happ H.H. and Wirgau K.A. (1982). Large scale optimal power flow. *IEEE PES Winter Meeting*, New York, Paper 82 WM 065-1.

19. Burchett R.C., Happ H.H. and Vierath D.R. (1984). Quadrically convergent optimal power flow. *IEEE PES Winter Meeting*, Paper 84 WM 045-1, Dallas, Texas.

20. Sun D.I., Ashley B., Brewer B. *et al.* (1984). Optimal power flow by Newton approach. *IEEE PES Winter Meeting*, Paper 84 WM 044-4, Dallas, Texas.

21. Aoki K. and Kanezashi M. (1985). A modified Newton method for optimal power flow using quadratic approximated power flow. *IEEE, PAS*-**104**(8), 2119–25.

22. Maria G.A. and Findlay J.A. (1987). A Newton optimal power flow program for Ontario Hydro EMS. *IEEE Trans. on Power Systems*, **2**(3), 576–84.

23. Franchi L., Innorta M. and Marannino P. (1983). The Han-Powell algorithm applied to the optimization of the reactive generation in a large-scale electric power system. *Proc. of the IFAC Symposium Large Scale Systems: Theory and Applications*, Warsaw, 178–84.

24. Giras T.C. and Talukdar S.N. (1981). Quasi-Newton method for optimal power flow. *Int. J. of Electrical Power and Energy Systems*, **3**(2), 59–64.

25. Bonnans J.F. (1989). Asymptotic admissibility of the unit stepsize in exact penalty methods. *SIAM J. on Control and Optimization*, **27**, 631–41.

26. Blanchon G., Bonnans J.F. and Dodu J.C. (1991). Application d'une méthode de programmation quadratique successive à l'optimisation des puissances dans les réseaux électriques de grande taille. *Bulletin de la Direction des Etudes et Recherches d'Electricité de France*, Série C(2), 67–101.

27. Blanchon G., Bonnans J.F. and Dodu J.C. (1990). Optimisation des réseaux électriques par une méthode de Newton. 9ième *Conférence Internationale Analyse et Optimisation des Systèmes*, Antibes, France, 1990. Also in *Lecture Notes in Information and Control Sciences*, 144, eds, A. Bensoussan and J.L. Lions, Springer-Verlag, Berlin, 423–31.

28. Blanchon G., Dodu J.C. and Léost J.Y. (1993). A Benders-type decomposition method for planning reactive power compensation devices. *Proc. of the 11th Power Systems Computation Conference*, Avignon, France, 419–27.

29. Blanchon G., Fourment C. and Mathieu E. (1993). Planning of reactive power compensation devices: QUASAR, a new optimization tool with security constraints. *IEEE-NTUA International Conference 'Athens Power Tech.'*, September '93, Athens, Greece.

30. Lebow W.M., Mehra R.K., Nadira R. *et al.* (1984). *Optimization of Reactive Volt-Ampere (VAR) Sources in System Planning. Volume 1: Solution Techniques, Computing Methods and Results.* EPRI EL-3729, Research Project 2109-1, Palo Alto.

31. Granville S., Pereira M.V.F. and Monticelli A. (1988). An integrated methodology for VAR sources planning. *IEEE Trans. on Power Systems*, **3**, 549–57, May.

32. Gomez T., Pérez-Arriaga I.J., Lumbreras J. and Parra V.M. (1991). A security-constrained decomposition approach for optimal reactive power planning. *IEEE Trans. on Power Systems*, **6**, 1069–76, August.

33. Geoffrion A.M. (1972). Generalized Benders decomposition. *J. of Optimization Theory and Applications*, **10**, 237–60.

34. Reid J.K. (1982). A sparsity exploiting variant of the Bartels–Golub decomposition for linear programming bases. *Mathematical Programming*, **24**, 55–69.

35. Subroutine LA05, *Harwell Subroutine Library* (1985). Computer Science and Systems Division, Harwell Laboratory, Oxfordshire, England.
36. Bertsekas D.P. (1982). *Constrained Optimization and Lagrange Multiplier Methods*, Academic Press, New York.
37. Bonnans J.F. and Gabay D. (1984). Une extension de la programmation quadratique successive. *Lecture Notes in Control and Information Sciences*, 63, Springer-Verlag, Berlin.
38. Han S.P. (1977). A globally convergent algorithm for nonlinear programming. *J. of Optimization Theory and Applications*, **22**, 297–309.
39. Mayne D.Q. and Maratos N. (1979). A first order exact penalty function algorithm for equality constrained optimization problems. *Mathematical Programming*, **16**, 303–24.
40. Mayne D.Q. and Polak E. (1982). A superlinearly convergent algorithm for constrained optimization problems, in *Mathematical Programming Study*, 16, North-Holland, 45–61.
41. Pschenichny B.N. (1970). Algorithms for the general problem of mathematical programming. *Kibernetika*, Kiev, **6**, 120–25.
42. Armijo L. (1966). Minimization of functions having continuous partial derivatives. *Pacific J. Math.*, **16**, p. 3.
43. Dodu J.C. and Huard P. (1988). Algorithme à pénalisation simple (méthode quadratique séquentielle non réalisable à convergence globale). *Rapport EDF-DER*, HR 30-1061, May.
44. Han S.P. (1976). Superlinearly convergent variable metric algorithms for general nonlinear programming problems. *Mathematical Programming*, **11**(3), 263–82.
45. Murray W. and Wright M.H. (1978). Projected Lagrangian methods based on the trajectories of penalty and barrier functions. *Systems Optimization Laboratory Report 78-23*, Standford University, California.
46. Powell M.J.D. (1978). The convergence of variable metric methods for nonlinearly constrained optimization calculations, in *Nonlinear Programming 3*, eds, O.L. Mangasarian, R.R. Meyer and S.M. Robinson, Academic Press, New York, 27–63.
47. Boggs P.T., Tolle J.W. and Wang Pyng (1982). On the local convergence of quasi-Newton methods for constrained optimization. *SIAM J. Cont. Opt.*, **20**, 161–71.
48. Coleman T.F. and Conn A.R. (1982). Nonlinear programming via an exact penalty function: asymptotic analysis. *Mathematical Programming*, **24**, 123–36.
49. Byrd R.H. (1985). An example of irregular convergence in some constrained optimization methods that use the projected Hessian. *Mathematical Programming*, **32**, 232–37.
50. Nocedal J. and Overton M.L. (1985). Projected Hessian updating algorithms for nonlinearly constrained optimization. *SIAM J. on Numerical Analysis*, **22**, 821–50.

51. Stoer J. and Tapia R.A. (1987). On the characterization of Q-superlinear convergence of quasi-Newton methods for constrained optimization. *Mathematics of Computation*, **49**, 581–84.

52. Dodu J.C. and Huard P. (1991). Quasi-Newton methods in nonlinear optimization. *EDF-DER Report*, HR 30-2329, July.

53. Dodu J.C. and Huard P. (1991). Utilisation de mises à jour doubles dans les méthodes de quasi-Newton. *Comptes rendus de l'Académie des Sciences Paris*, **313**, series I, 329–34.

54. Broyden C.G., Dennis J.E. and Moré J.J. (1973). On the local and superlinear convergence of quasi-Newton methods. *J. Inst. Maths. Applics*, **12**, 223–45.

55. Dennis J.E. and Moré J.J. (1973). A characterization of superlinear convergence and its application to quasi-Newton methods. *Mathematics of Computation*, **28**(126), 549–60.

56. Dennis J.E. and Moré J.J. (1974). Quasi-Newton methods, motivation and theory. *SIAM Review*, **19**(1), 46–89.

4

SHORT-CIRCUIT CURRENTS

4.1 INTRODUCTION

The currents and voltages which appear during faults (short-circuits or insulation faults) affecting electrical systems play an important role in their operation.

In fact:

- they are fundamental parameters in the design of the equipment, particularly the circuit breakers whose task is to interrupt fault currents;
- they are the input variables of the protection equipment for fault detection and location;
- fault currents to earth induce currents in neighbouring conductive circuits, such as metallic piping and telecommunications circuits, which can affect their correct operation and safety.

Different approaches can be used depending on the application envisaged.

When determining the rating of the equipment, we try to obtain a maximum short-circuit value quickly. In general, this value corresponds to three-phase solid short-circuits at the nodes which require consideration only of the system of positive components.

In this context, the following approximations are made:

- The conditions before the fault are not taken into account: the voltage map is assumed to be uniform (constant moduli, zero phases) over the whole network. The power flow currents are therefore ignored in the presence of the fault currents.

58 SHORT-CIRCUIT CURRENTS

- The resistances and susceptances of equipments are ignored, as are also the loads.

- Transformers are assumed to be operating on their rated taps.

- To avoid underestimating the short-circuit currents, it is assumed that all generating units are in service and the network is complete (minimum impedances).

Under these conditions, corresponding to extreme hypotheses, the equipment on the system can be oversized.

With a view to adopting more realistic values, a probabilistic approach can be used (Monte-Carlo method) to determine the short-circuit currents, permitting evaluation of the sensitivity to operating hypotheses: the structure of the power system and starting-up of generating sets.

Many situations of availability of structures such as lines, transformers and generating sets are thus considered taken at random. These situations are used as data for calculating short-circuit currents. One can then determine the frequency of occurrence of the extreme values of these currents.

When an accurate evaluation corresponding to clearly defined hypotheses is sought, more rigorous modelling must be applied.

4.2 DEFINITION OF THE SHORT-CIRCUIT CURRENT

When a short-circuit appears, on account of the inductive nature of the network, the value at power frequency is not immediately established. The current can be considered approximately as the sum of two terms (Figure 4.1):

- an aperiodic transient component which decreases rapidly in the form $I\sqrt{2}\, e^{-t/T} \cos\theta$ – the time constant T is of the order of 50 to 100 ms for an extremely high voltage system, depending on the characteristics of the system and the location of the fault, and θ is the angular deviation between the zero of the voltage at power frequency and the instant at which the short-circuit appears;

- a sinusoidal component at power frequency corresponding to the steady state in the form $I\sqrt{2}\cos(\omega t + \theta)$. When the fault is electrically close to machines, the variation in the internal impedance of the machines with time causes the amplitude of the short-circuit current to

DEFINITION OF THE SHORT-CIRCUIT CURRENT 59

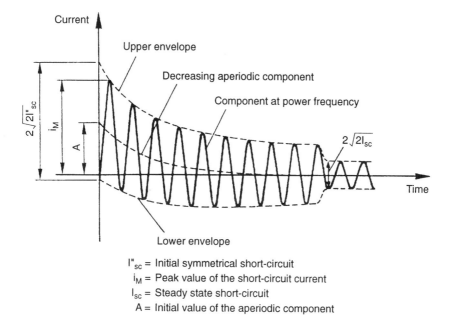

I''_{sc} = Initial symmetrical short-circuit
i_M = Peak value of the short-circuit current
I_{sc} = Steady state short-circuit
A = Initial value of the aperiodic component

Fig. 4.1   The general pattern of a short-circuit current.

vary also with time. The internal impedance of the machines should therefore be selected as a function of the moment considered, as indicated in section 4.4.3. If the fault is very close to the machines, it is no longer possible to ignore the effect of the voltage regulators (overexcitation). For this special case, it is necessary to consider detailed modelling of the generators, such as that used for simulating stability phenomena.

Moreover, the discontinuity caused by the appearance of the short-circuit generates transients which are rapidly damped (in a few milliseconds) and are not taken into account by classic short-circuit current calculations. If we wish to take into account these phenomena, simulation of very high speed transients must be used, like that considered in Chapter 7. This is the case particularly for the investigations concerning protections with a very short response time operating during these transient phenomena.

4.3 METHOD OF CALCULATING SHORT-CIRCUIT CURRENTS

4.3.1 Method of symmetrical components applied to the calculation of short-circuit currents

In the usual calculations of short-circuit currents, it is assumed that:

- the high speed transient components have disappeared;

- the mechanical variables of the machines have not had time to change on account of their inertia and the fact that the voltage regulating devices have not had time to act. The generating sets are therefore represented by a constant e.m.f. behind a given reactance as a function of the moment considered as indicated in section 4.4.3.

Let us consider a symmetrical network supplied with power by generators with balanced e.m.f. values, where there are local imbalances caused by the presence of unequal impedances on the three phases. This is the case particularly for unsymmetrical faults. We seek to determine the currents corresponding to the fault location and the currents and voltage on the whole of the system.

The short-circuit current calculations are based on the application of Thevenin's theorem, which is developed from the principle of superimposing, applicable to linear networks (Figure 4.2).

If an external element of impedance Z_e is connected between any two points M and N of an active linear network, the current I flowing through this impedance is calculated by considering the network between points M and N as similar to a single source, the e.m.f. of which would be equal to the pre-existing voltage V_p and the internal impedance equal to the impedance Z_p of the network made passive, seen from points M and N.

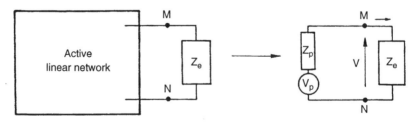

Fig. 4.2 Thevenin's theorem.

METHOD OF CALCULATING SHORT-CIRCUIT CURRENTS

$$V = Z_e I = V_p - Z_p I; \quad \text{that is } I = \frac{V_p}{Z_e + Z_p} \tag{4.3.1}$$

The method consists of:

- isolating the unbalanced zone from the remainder of the network, reducing it as far as possible;
- applying the method of symmetrical components to the currents and voltages in the symmetrical part of the network outside this zone;
- applying the classic formulae, using phase currents and voltages, inside the isolated zone;
- expressing the continuity of the electrical variables of voltages and currents at the boundary delimiting the zone;
- solving the system of equations permitting the calculation of electrical variables at all points of the network.

Let us examine the principle of the calculating method taking a simple example.

Calculation of the fault current in the case of a short-circuit at one point on the network

The fault can be considered as a pinpoint dissymmetry separated from the network (which retains a symmetrical configuration) by an imaginary boundary (Figure 4.3).

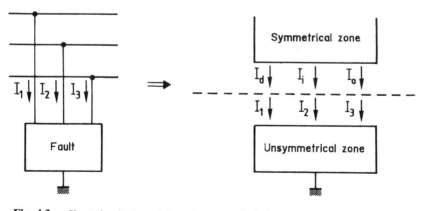

Fig. 4.3   Short-circuit at a point on the system: isolation of the unsymmetrical zone.

62 SHORT-CIRCUIT CURRENTS

In the unsymmetrical zone, the fault is expressed in phase components.

For example: three-phase fault: $V_1 = V_2 = V_3 = 0$
single-phase fault: $V_1 = 0; I_2 = I_3 = 0$

The symmetrical zone can be broken down into three equivalent single-phase circuit diagrams, and by applying Thevenin's theorem to the symmetrical components associated with each diagram, we obtain three equations:

$$\begin{cases} V_d = V_d^p - Z_d I_d \\ V_i = 0 - Z_i I_i \\ V_0 = 0 - Z_0 I_0 \end{cases} \quad (4.3.2)$$

in which V_d^p is the pre-existing voltage at fault point Z_d, and Z_i and Z_0 are the impedances equivalent to the network in the three systems.

By considering the continuity of the values at the boundary, we convert the phase variables described in the unsymmetrical zone into symmetrical variables.

The solution of the system with six equations and six unknowns makes it possible to obtain the currents and voltages at point M, as symmetrical coordinates, and then as phase coordinates.

To illustrate this method, we shall consider as an example the case of a three-phase and then single-phase solid short-circuit.

Three-phase fault

Expression of the fault: $\quad V_1 = V_2 = V_3 = 0$

Conversion into symmetrical components: $V_d = V_i = V_0 = 0$

which is, according to (4.3.2): $\quad I_d = \dfrac{V_d^p}{Z_d}; I_i = I_0 = 0$

Return to the phase variables: $\quad V_1 = V_2 = V_3 = 0$

$$I_1 = \frac{V_d^p}{Z_d}; I_2 = \frac{a^2 V_d^p}{Z_d}; I_3 = \frac{a V_d^p}{Z_d}$$

Single-phase fault on phase 1

Expression of the fault: $\quad V_1 = 0; I_2 = I_3 = 0$

Conversion into symmetrical components: $V_d + V_i + V_0 = 0$
$I_d = I_i = I_0$

which is, according to (4.3.2):
$$I_d = I_i = I_0 = \frac{V_d^p}{Z_d + Z_i + Z_0}$$

$$V_d = \frac{Z_i + Z_0}{Z_d + Z_i + Z_0} V_d^p; \quad V_i = -\frac{Z_i}{Z_d + Z_i + Z_0} V_d^p; \quad V_0 = -\frac{Z_0}{Z_d + Z_i + Z_0} V_d^p$$

Return to phase variables:
$V_1 = 0$

$$V_2 = \frac{(a^2 - a)Z_i + (a^2 - 1)Z_0}{Z_d + Z_i + Z_0} V_d^p$$

$$V_3 = \frac{(a - a^2)Z_i + (a - 1)Z_0}{Z_d + Z_i + Z_0} V_d^p \quad (4.3.3)$$

$$I_1 = \frac{3V_d^p}{Z_d + Z_i + Z_0}; \quad I_2 = I_3 = 0$$

where a is the operator which defines a phase shift of 120°, and which is described by

$a = e^{j120°}$

4.3.2 Calculation of electrical variables in the general case

The general method explained previously is used, proceeding as follows:

- The zone of dissymmetry is located and isolated from the remainder of the network. Here one may be led to consider a new topology (for example if the fault consists of a short-circuit preceded by single-phase or three-phase opening of lines). As the boundary is based on m nodes, writing the fault characteristics in phase variables provides $3 \times m$ equations which are converted into symmetrical components.

- The zone of dissymmetry being disconnected, the remainder of the network, which has a symmetrical configuration, can be broken down into three single-phase equivalent circuits: positive, negative and zero sequences.

In each of these circuits, the currents and voltages are linked by Ohm's and Kirchhoff's laws which can be written in the form of a matrix equation:

64 SHORT-CIRCUIT CURRENTS

$$I = I_p - YV \qquad (4.3.4)$$

where:
- I Vector of the fault current injections at the nodes (zero elements except at m boundary nodes).
- I_p Vector of the current injections before the fault (in general E/jX'_d) at the nodes (zero elements except at the nodes onto which the generating sets deliver power). This vector is zero in negative and zero systems.
- Y Nodal admittance matrix in short-circuit.
- V Vector of the voltages on all the nodes of the symmetrical zone.

In the absence of a fault ($I = 0$), I_p created the pre-existing voltage map V_p, hence: $I_p = YV_p$.

The equation can therefore be written:

$$I = Y(V_p - V)$$
that is: $V = V_p - ZI$ (4.3.5)
with: $Z = Y^{-1}$

in the positive, negative and zero systems.

By extracting from (4.3.5) the m lines corresponding to the m boundary nodes, we obtain m relationships in each system:

Positive system $\quad V_{d_f} = V_d^p - Z_{d_{ff}} I_{d_f}$
Negative system $\quad V_{i_f} = 0_f - Z_{i_{ff}} I_{i_f}$
Zero system $\quad V_{0_f} = 0_f - Z_{0_{ff}} I_{0_f}$ (4.3.6)

These relationships, which reflect Thevenin's theorem, generalized to m nodes, provide $3 \times m$ equations.

Assuming that the pre-existing voltage map is known (calculated or obtained from approximations), solving the $3 \times m$ fault equations (4.3.3) and $3 \times m$ equations (4.3.6) provides the $3 \times m$ currents I_f and $3 \times m$ voltages V_f at the m boundary nodes.

The voltages V can then be determined in symmetrical variables, and then in phase variables, on all the nodes of the symmetrical zone, by solving (4.3.5): $V = V_p - ZI$ in which the only elements of I which are not zero constitute I_f.

From this we deduce the currents in all the elements of the network, by applying Ohm's law.

4.3.3 Techniques of calculation

A short-circuit current calculation always comes down to a (total or partial) solution of linear systems.

For each fault configuration, we have to obtain the pre-existing voltage map and certain elements of the matrices Z_d, Z_i, Z_0.

Depending on the type of fault studied (very simple fault such as a three-phase short-circuit on an electrical node, or more complex, such as a fault between the phases of two circuits of a double line) and the type of results desired (short-circuit power on all the nodes or values on the whole network for a particular fault), different techniques are possible.

Generally speaking, the procedure is as follows:

a) Calculation of the values before considering the faults

- The pre-existing voltage map is determined: the exact voltage map (moduli and phases) obtained from an active and reactive power flow calculation, or obtained by approximation.

- The matrices Y_d, Y_i and Y_0 are constructed. These matrices are symmetrical and have complex elements. The positive and negative matrices are very sparse and are often considered identical. The zero-sequence matrix has a different topology and values on account of the transformers, of the neutral earth, and possible coupling by mutual impedance between lines on shared supports.

- The negative matrices Z_d, Z_i and Z_0 are not calculated explicitly (these matrices are full), but the matrices Y are factorized in the form of a lower triangular matrix and an upper triangular matrix:

$$Y = LU \qquad (4.3.7)$$

It will then be quick and easy to calculate the useful columns of the impedance matrices, and to calculate and store certain key elements of these matrices forming the 'sparse negative'.

b) Taking faults into account

Using these basic data, methods of compensation make it possible to simulate any changes in topology associated with faults and to determine the new associated pre-existing voltage maps.

Examples:

Fig. 4.4 Single fault at one node.

Single fault at one node. No change in topology. Rapid calculation of the column $Z_{\cdot M}$.

Direct use of:

$$V = V_p - ZI$$

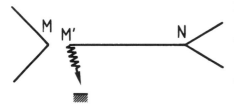

Fig. 4.5 Single fault at the end of an open line.

Fault at the end of an open line. Change in topology. Calculation of columns $Z_{\cdot M}$ and $Z'_{\cdot M}$ and voltages V'_p starting from Z and V_p.

Then use $V = V'_p - Z'I$

4.4 MODELLING NETWORK ELEMENTS FOR SHORT-CIRCUIT CALCULATIONS

4.4.1 Overhead lines and cables

The modelling is similar to that explained in Chapter 3 for the load flow calculations. However, shunt admittances whose currents are negligible in relation to those of the short circuit are not taken into account. One can also ignore the resistance in relation to the reactance of the series impedance providing that

$$\text{arc tan tg}\left(\frac{L\omega}{R}\right)$$

remains greater than approximately 60°, which is the case for transmission lines for voltages higher than approximately 30 kV. For lines with lower voltage, the relevance of such a simplification should be examined.

4.2.2 Transformers

The modelling of Chapter 3 for load flow calculations is still valid. For short-circuit current calculations, the magnetizing impedances are ignored.

In the positive and negative systems, the transformers are modelled by their short-circuit impedance, the resistance of which is ignored in view of the reactance.

For the zero-phase sequence system, the reactance is dependent on the mode of coupling the windings, the nature of the magnetic circuit, and the earthing of the neutral.

For load flow calculations, as a first approach, a representation is adopted, which strictly is applicable only to transformers with separate single-phase elements, which is generally a sufficient approximation.

For the short-circuit current calculations, one may need to seek greater accuracy, particularly for the protection settings. The main cases of connection of windings and of the magnetic circuit will be found below.

4.4.2.1   Star-delta transformer

The zero-phase sequence reactance seen from the terminals of the delta winding is infinite, as zero-phase sequence currents cannot flow since there is no return path; the same applies on the star side when the neutral is not connected to earth.

The zero-phase sequence reactance seen from the terminals of the star winding when the neutral is connected to earth is more or less equal to the positive (or negative) reactance of the transformer, whether the latter has a magnetic circuit with free flux (4 or 5 cores) or with forced flux (3 cores).

If the winding which has its neutral earthed is in zigzag connection, the zero-phase sequence reactance corresponds to the leakage reactance between the two zigzag half-windings. This reactance is very low, and its value is approximately 1%.

The star-delta transformer, with the neutral earthed, makes it possible to achieve perfect decoupling between two networks, between the zero-sequence components of the currents and voltages (Figure 4.6).

In fact, the currents in phase I_{01}, flowing in the phases of the primary, induce equal currents I_{02} which circulate in the delta-connected secondary windings.

In a secondary winding, the voltage drop caused by the passage of the current I_{02} exactly balances the voltage induced by the primary winding, from which it follows that the voltage at the three terminals A, B and C of the secondary is zero. From the point of view of zero-sequence components, the fact of connecting or disconnecting a network R does not change the operating conditions of the transformer considered and consequently the value of the zero-phase sequence impedance seen from the primary.

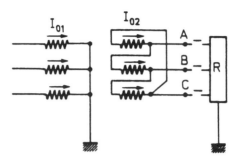

Fig. 4.6   Star-delta transformer.

4.2.2.2   Star-star transformer

The two neutral points are connected to earth or to a neutral conductor

The zero-phase sequence reactance of a star-star transformer is, under these conditions, strictly equal to its positive reactance if the magnetic circuit has separate cores, and more or less equal if the magnetic circuit has free fluxes (4 or 5 cores) and it differs only slightly if the magnetic circuit is of the forced flux type.

Only one of the neutral points is connected to earth

The zero-phase sequence reactance seen from the terminals of the winding whose neutral is insulated is infinite.

If the transformer has free flux or separate magnetic circuits, it is presented as a three-phase inductance coil whose zero-phase sequence reactance is equal to the no-load positive reactance; it is therefore very high.

The different possible circuits are shown on Figure 4.7.

4.4.2.3   Reactances of transformers with three windings

The transformer with three windings permits power exchanges between three power systems with different voltage levels.

Positive and negative reactances

It is demonstrated that a three-phase transformer with three windings can be represented by a single-phase equivalent circuit with a star having three

NETWORK ELEMENTS 69

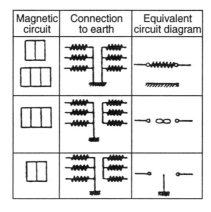

Fig. 4.7   Star-star transformer.

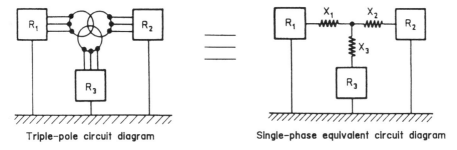

Triple-pole circuit diagram   Single-phase equivalent circuit diagram

Fig. 4.8   Transformer with three windings: triple-pole and single-phase equivalent circuit diagrams.

branches linking together the networks envisaged. The values X_1, X_2 and X_3 of the reactances of the three branches of the star are identical in the positive system and in the negative system (Figure 4.8).

The elements X_1, X_2 and X_3 of the imaginary star can be determined very simply if the results of three short-circuit tests carried out on the windings, taking two at a time, are known.

Let X_{12} be the short-circuit reactance measured between windings 1 and 2, winding 2 being short-circuited and winding 3 open.

Let X_{13} and X_{23} be the short-circuit reactances of the other windings, taken two at a time.

It is demonstrated that:

$X_{12} = X_1 + X_2$
$X_{23} = X_2 + X_3$
$X_{31} = X_3 + X_1$

the reactances of the three branches of the star are calculated by relating them to the rated output and the rated voltage of one of the windings, which is most often the high voltage. This method of representation is valid whatever the form of coupling of the windings constituting the transformer (delta-star or zigzag).

The star circuit with three branches being imaginary, often one of the branches has a negative impedance (generally low); so there is no need to find its physical significance.

Zero-phase sequence reactance

The zero-phase sequence reactance is dependent on the connection of the windings.

Figure 4.9 shows a star-delta-star transformer with only one of the neutral points of the transformer connected to earth. The zero-phase sequence flowing through each of the phases of star 1 induces currents which circulate in the delta 3 and the zero-phase sequence reactance, seen from terminal 1, is equal to the leakage reactance between the star and delta windings. The zero-phase sequence does not therefore cross the transformer.

If Z_1 and Z_3 are the positive reactances of branches 1 and 3 of the star, the zero-phase sequence reactance is equal to:

$$X_0 = X_1 + X_3$$

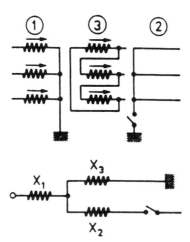

Fig. 4.9   Star-delta-star transformer: zero-phase sequence circuit; one neutral point grounded.

NETWORK ELEMENTS 71

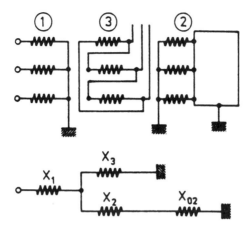

Fig. 4.10 Star-delta-transformer: zero-phase sequence circuit; two neutral points grounded.

Seen from terminals 2 and 3, the zero-phase sequence reactance of the transformer is evidently infinite.

Figure 4.10 shows a star-delta-star transformer with two neutral points connected to earth. The zero-phase sequence of the currents can cross the transformer if there is at least one neutral in the network supplied with power by winding 2.

Let us assume that winding 2 is connected onto a network with polar reactance X_{02}, we demonstrate that the zero-phase sequence reactance of the transformer and network assembly seen from the terminals of winding 1 is equal to:

$$X_0 = X_1 + \frac{X_3(X_2 + X_{02})}{X_3 + X_2 + X_{02}}$$

Seen from winding 3, the zero-phase sequence reactance is infinite; reasoning identical to that above shows that the zero-phase sequence reactance seen from winding 2 is equal to:

$$X_2 + \frac{X_3(X_1 + X_{01})}{X_3 + X_1 + X_{01}}$$

Figure 4.11 shows a star-delta-star transformer with neutral earthed. The zero-phase sequence currents can flow in the primary windings. No zero-phase sequence power is exchanged between network 1 and net-

72 SHORT-CIRCUIT CURRENTS

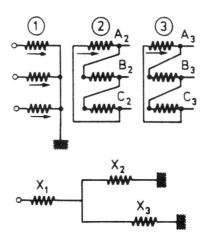

Fig. 4.11 Star-delta-star transformer: zero-phase sequence circuit seen from primary; neutral point grounded.

works 2 and 3 since the voltages at points such as A_2 B_2 C_2 and A_3 B_3 C_3 are zero.

Currents flow in the delta connections 2 and 3 but do not penetrate onto networks 2 and 3. Seen from the primary, the equivalent circuit of the transformer is reduced to the star connection with three branches X_1, X_2 and X_3; the reactances have their ends connected to earth.

The zero-phase sequence reactance of the transformer, seen from the primary, is under these conditions:

$$X_0 = X_1 + \frac{X_2 X_3}{X_2 X_3}$$

Seen from the secondary or the tertiary, the zero-phase sequence reactance of the transformer is obviously infinite.

Figure 4.12 shows a transformer with a zigzag winding: star delta, zigzag, two neutrals connected to earth. The zero-phase sequence currents flow in the primary. They induce currents in the secondary delta winding but these currents circulate and do not appear in network 2.

No zero-phase sequence current is induced in the zigzag winding by the star and delta windings as the voltages induced in the two half-coils of a phase cancel one another out, as the windings go in opposite directions.

The zero-phase sequence impedance of the transformer, seen from the primary, is:

$X_0 = X_1 + X_2$

Conversely, in the transformer shown in Figure 4.13, the zero-phase sequence currents coming from the zigzag tertiary do not create any flux in the transformer. There cannot therefore be any zero-phase sequence power exchange.

The zero-phase sequence impedance of the tertiary is very low since the currents are limited only by the resistances of the windings and the magnetic leakage between two coaxial coils of the zigzag winding. The currents can adopt high values if the reactance X_{03} of the remainder of network 3 itself is low.

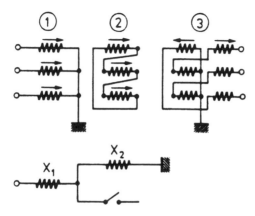

Fig. 4.12 Star-delta-zigzag transformer: zero-phase sequence circuit seen from primary; two neutral points grounded.

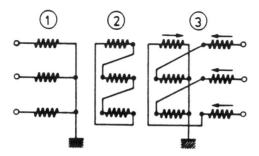

Fig. 4.13 Star-delta-zigzag transformer: zero-phase sequence seen from tertiary; two neutral points grounded.

74 SHORT-CIRCUIT CURRENTS

To summarize, the zero-phase sequence reactance seen from the primary is:

$$X_0 = X_1 + X_2$$

The zero-phase sequence reactance seen from the secondary is infinite.

The zero-phase sequence reactance seen from the tertiary is close to 1% of the rated power S_n of the transformer.

The values of the zero-phase sequence resistances of the transformers with two or three windings, depending on their form of connection, are summarized in Table 4.1.

4.4.3 Synchronous machines

The modelling of synchronous machines for short-circuit current calculations is similar to that explained in Chapter 3 for power flow calculations. However, the electromagnetic phenomena involved in a short-circuit close to the machine cause the internal impedances to vary with the moment of time considered. In the following text, the resistances of these impedances in relation to the reactances will be ignored.

Positive sequence

The machine is represented by an e.m.f. behind a reactance. In the first instants following the appearance of the short-circuit ($t < 0.10$ s) the reactance to be taken into account is the subtransient reactance. For approximately 0.10 s $< t < 1$ s, we take the transient reactance. Then, for approximately $t > 1$ s, we can assume that the transient conditions are sufficiently damped and take the synchronous reactance of the steady state.

At present, faults are normally cleared in less than 1 s. When designing the equipment and adjusting the protections, the transient reactance is therefore considered in the short-circuit calculations.

The value of synchronous, transient and subtransient reactances varies according to the type of machine, with non-salient or salient poles.

Negative sequence

The electrical variables of this system correspond to a field rotating in the opposite direction to the positive system. The negative-phase sequence

NETWORK ELEMENTS 75

Table 4.1 Zero-phase sequence reactance of transformers; equivalent circuit diagrams; reactance value

Group			Single line equivalent diagram	Value of the zero-phase sequence reactance of the transformer, seen		
Primary	Secondary	Tertiary		From the primary terminals 1	From the secondary terminals 2	From the tertiary terminals 3
			1 —o 2 o—	Infinite	Infinite	
			1 —o 2 o—	Infinite	Infinite	
			1 X_{11} 2 / 1 2	Free fluxes infinite Forced fluxes X_{11} = 10–15 times X_{cc}	F.L. infinite F.F. infinite	
			1 —⋀⋀⋀— 2	$X_{12} = X_{cc}$	$X_{12} = X_{cc}$	
			—o o—	Infinite	Infinite	
			1 X_{12} 2	$X_{12} = X_{cc}$	Infinite	
			—o o—	Infinite	Infinite	
			X_{22} 1 —o—⋀⋀— 2	Infinite	X_{22} = 1% of S_n	
			1 X_{11} 2 / 1 2	F.L. infinite F.F. X_{11} = 10–15 times X_{cc}	F.L. infinite F.F. infinite	

F.L. = free flux
F.F. = forced flux

76 SHORT-CIRCUIT CURRENTS

Table 4.1 Cont'd

Group			Single line equivalent diagram	Value of the zero-phase sequence reactance of the transformer, seen		
Primary	Secondary	Tertiary		From the primary terminals 1	From the secondary terminals 2	From the tertiary terminals 3
(diagram)	(diagram)		X_{22} diagram	Infinite	$X_{22} = 1\%$	
(diagram)	(diagram)		X_{22} diagram	Free fluxes infinite; Forced fluxes $X_{11} = 10\text{--}15$ times X_{cc}	Free fluxes X_{22} 1% S_n; $X_{22} = 1\%\ S_n$	
(diagram)	(diagram)		(diagram)	Infinite	Infinite	
(diagram)	(diagram)	(diagram)	X_{11} diagram	Free fluxes infinite; Forced fluxes $X_{11} = 10\text{--}15$ times X_{12}	Infinite	Infinite
(diagram)	(diagram)	(diagram)	$X_{01}, X_1, X_2, X_{02}, X_3, X_{03}$ diagram	$X_1 + \dfrac{(X_2+X_{02})(X_3+X_{03})}{X_2+X_3+X_{02}+X_{03}}$	$X_2 + \dfrac{(X_1+X_{01})(X_3+X_{03})}{X_1+X_3+X_{01}+X_{03}}$	$X_3 + \dfrac{(X_1+X_{01})(X_2+X_{02})}{X_1+X_2+X_{01}+X_{02}}$
(diagram)	(diagram)	(diagram)	X_1, X_2, X_3 diagram	$X_1 + \dfrac{X_2 X_3}{X_2+X_3}$	Infinite	Infinite
(diagram)	(diagram)	(diagram)	$X_{01}, X_1, X_2, X_3, X_{03}$ diagram	$X_1 + \dfrac{X_2(X_3+X_{03})}{X_2+X_3+X_{03}}$	Infinite	$X_1 + \dfrac{X_3(X_1+X_{01})}{X_1+X_2+X_{01}}$
(diagram)	(diagram)	(diagram)	X_1, X_2, X_{33} diagram	$X_1 + X_2 = X_{12}$	Infinite	$X_{33} = 1\%\ S_n$

reactance corresponds to the flux which rotates in relation to the inductor at twice the synchronous speed. It is generally weaker than the subtransient reactance (by 5–30%), but it can be taken as equal to the latter in the absence of accurate information.

Zero-phase sequence

The zero-phase sequence reactance is generally infinite as synchronous machines are normally connected to the network by transformers whose windings, on the machine side, are delta-connected or star-connected with insulated neutral.

In the case of a machine with its neutral point connected to earth, the zero-phase sequence reactance is always very low (20–80% of the negative phase sequence reactance) and should be considered only if the neutral of the alternator is connected to earth directly or via a very low impedance.

4.4.4 Asynchronous machines

During a short-circuit close to an asynchronous machine, the transient state of the current is damped very quickly (in one or two periods), and the machine then behaves like a passive impedance with the value U_n^2/S_n (where U_n is rated voltage between phases; S_n is apparent rated output).

This is the value generally adopted for short-circuit current calculations if there is no particular problem, for the positive and negative components of the internal reactance.

As for synchronous machines, the zero-phase sequence reactance represents only a small fraction of the positive phase sequence reactance. Moreover, this reactance is not taken into account in calculations of the short-circuit current, as asynchronous machines are installed with the neutral point insulated.

4.4.5 Loads

The modelling is similar to that of Chapter 3 for the power flow calculations: impedance to earth identical for the positive and negative systems. The zero-phase sequence impedance is not to be taken into account, the loads always being supplied with power by transformers with the neutral insulated on the network side.

Depending on the aim of the research, in certain short-circuit investigations, the loads can be ignored.

FURTHER READING

Alsac O., Scott B. and Tinney W.F. (1983). Sparsity-oriented compensation methods for modified network solution. *IEEE*, **PAS-102**(5), May.

Anderson P.M. (1973). *Analysis of Faulted Power Systems*, Iowa State University Press.

Back H., Le Saout V. and Meslier F. (1988). A new computerized system to help planning the power transmission networks, *IEEE/PES, Summer Meeting*, Portland.

Brandwajn V. and Tinney W.F. (1985). Generalized method for fault analysis. *IEEE*, **PAS-104**(6), June.

Davriu A., Giard A. and Verseille J. (1990). Tristan: A probabilistic tool for the analysis of short circuit currents: Methodology and examples of applications. Power Systems Computing Conference, Graz.

El-Kady M.A. (1983). Probabilistic short-circuit analysis by Monte Carlo simulation. *IEEE*, **PAS-102**(5) May.

El-Kady M.A. and Ford G.L. (1983). An advanced probabilistic short-circuit program. *IEEE*, **PAS-102**(5) May.

Han Z.X. (1982). Generalized methods of analysis of simultaneous faults in electric power system, *IEEE*, **PAS-101**(10) Oct.

Pelissier R. (1971). Les réseaux d'énergie électrique, Editions Dunod, Paris.

Pelissier, R. (1977). La puissance de court-circuit dans les réseaux de transport et de distribution d'énergie électrique, *Revue Générale de l'Electricité*, **86**(2), Feb.

Takamasmi K., Fagan J. and Chen M.S. (1973). Formation of a sparse bus impedance matrix and its application to short circuit study. *PICA*, Minneapolis, 3–6 June.

5

LONG-TERM DYNAMICS

5.1 INTRODUCTION

In the preceding chapters, we described the techniques used to investigate steady states of the electrical system. We shall now consider dynamic phenomena associated with maintaining the power system in a condition of equilibrium by the action of continuous closed-loop control systems based on macroscopic actions.

Essentially this involves analysis of the behaviour of the voltage and maintaining the balance between generation and consumption, that is, control of the frequency and power flows on international interconnections. For this purpose, it is necessary to monitor the changing pattern of different electrical or mechanical variables with time, over several minutes, tens of minutes or even several hours.

The simplest model which might be suitable would involve calculating active and reactive power flows after each change in the topology of the network, the power generated and the consumption.

Reactive power generated by the generating sets should be checked to ensure that it remains within permissible ranges. If this limit is reached by a generating set, it is its reactive output which should be kept to a value equal to the limit reached, and no longer its voltage. The calculation should be continued with this hypothesis.

To represent the taking on of load or a loss of generation, the load flow calculation would be modified to take into account:

- the final conditions of the primary speed control action on generating sets (introduction of governor droop) or secondary power frequency control, if present (i.e. introduction of the control level);
- the on-load tap changers of the transformers;
- the final conditions of the secondary voltage control, if present.

However, such a model does not take into account slow transient phenomena. A correct state can be found although, taking into account time constants, thresholds and response delay times of the various elements of the system, one would observe transient states on the actual power system which could cause the equipments to react, changing the evolution of the system accordingly, such as:

- overload tripping of elements;
- load shedding or islanding when frequency falls;
- tripping out of generating sets by auxiliary protections based on voltage or frequency drop.

To represent the changes in the system correctly, taking into account slow transient phenomena, a model best suited to the simulation of long term dynamic phenomena can be used.

5.2 LONG-TERM DYNAMICS MODEL

5.2.1 General principles

The long-term dynamics model brings a great improvement in the calculation of the frequency, reflecting the dynamic balance between generation and consumption.

The frequency is considered equal at any point on the power system. In view of their frequency (of the order of 1 Hz), oscillations between the rotors of the different machines which have the same speed and the same acceleration are ignored (tied rotors hypothesis).

We add to the load flow calculation equations concerning the network, differential equations representing the dynamic operation of the turbine and boiler generating assemblies, mechanical equations for shaft lines and local or centralized controls which interact with the phenomena simulated.

5.2.2 Method of calculation

The classic organisation of a tool for simulating long-term dynamics is as follows:

- At time t, the set of equations of the turbine and items upstream of the turbine will be integrated with an interval h for each generating set. Starting at the state with the preceding time interval, we thus gain access to the mechanical power of each set, as an explicit function of the internal variables and deviation in relation to the reference frequency.

- By assuming that the overall consumption of the electrical system is constant throughout the time interval h, we shall reach the common acceleration of the rotors (and hence the frequency) by the dynamic equation

$$\sum_{\text{sets}} P_{\text{mech}} - \sum_{\text{sets}} P_{\text{elec}} = I\omega\gamma$$

and the integration of the equation

$$\gamma = \frac{d\omega}{dt}$$

where I is the total moment of inertia of the connected subnetwork under review, and ω is the electrical angular frequency; the electrical power considered is that of the preceding interval.

- Every k time intervals (therefore with an interval $H = kh$), an active-reactive load flow calculation is carried out and the unit electric power values of the generating sets $P_{\text{elec}\,i}$ are updated by solving, for each generating set i, the mechanical equation

$$P_{\text{mech}\,i} - P_{\text{elec}\,i} = I_i \omega \gamma$$

where I_i designates the inertia of the set i.

- For the network, the system of nonlinear equations is

$$\begin{cases} P_{\text{elec}\,i} = \varphi_i(\boldsymbol{\theta}, V) \\ Q_i = \Psi_i(\boldsymbol{\theta}, V) \end{cases} \text{(classic power flow, cf. Chapter 3)}$$

to which we add the equations arising from the integration of differential equations modelling the turbine and items upstream of the turbine, and the frequency equation, as for interval h.

This problem takes the following form:

$$\begin{cases} P_i = \varphi_i(\boldsymbol{\theta}, V, \gamma) \\ Q_i = \Psi_i(\boldsymbol{\theta}, V) \end{cases}$$

82 LONG-TERM DYNAMICS

This system is solved by a classic Newton–Raphson method.

Starting with the initial values θ, V and γ, we calculate the deviations ΔP and ΔQ, between the desired values and the calculated values of power.

We solve the system:

$$\begin{pmatrix} \Delta P \\ \Delta Q \end{pmatrix} = \mathbf{J} \begin{pmatrix} \Delta \theta \\ \Delta V \\ \Delta \gamma \end{pmatrix}$$

We correct θ, V and γ and iterate until $\{|\Delta P|, |\Delta Q|\}$ is sufficiently small.

The Jacobian $\mathbf{J}$ is updated at each iteration, taking into account the new values of V and Q. In this type of problem, the decoupled Jacobian technique cannot be used. The complete Jacobian must be considered.

The terms $\partial P/\partial V$ and $\partial Q/\partial \theta$ become predominant near to voltage collapse, leading to divergence of the calculation. At the present time this phenomenon is one of the main problems of long-term dynamics of power systems.

At each time interval, we check that all the quantities calculated are indeed in the operating ranges envisaged (if this is not the case, we take into account the limitations or protections which may for example lead to the tripping of a generating set). The variation in consumption according to the load variation function which we have adopted (depending of the investigation to be carried out) is also calculated, and the operation of local automatic controls (for example on-load tap changers) is taken into account.

The general problem of simulating long-term dynamics has thus been divided into three subproblems which are easier to solve:

- The integration of nonlinear differential equations, with a short time interval (500 ms for example), typically using the trapezoid method in order to promote numerical stability, coupled with solving the overall dynamic equation (trapezoidal or Euler) with fixed active total consumption.

- Every k time intervals (therefore with an interval $H = kh$), a calculation load flow on the system as a whole, coupled with the integration of the differential equations of the generating sets. This is a nonlinear problem similar to that of Chapter 3. It is solved with the techniques described above, giving priority to good convergence to the detriment of the calculation time, in order to be able to study the electrical states close to instability, that is the singularity of the Jacobian.

- Taking into account the events affecting the systems, the change in state of the local automatic controls, and the calculation of the command transmitted by the centralized controls.

This decoupling permits good efficiency in terms of calculating time (Appendix 5.A). Simulators for training operating personnel can thus be created, in which the dynamic changes in a large electrical system are represented (Chapter 9).

Many studies have been undertaken into the numerical stability of such methods. They enable the users to find suitable values for the calculation parameters (h,k) fairly easily. However, we shall see further on that the tools for investigating long-term dynamics have disadvantages which have led to the development of new methods.

5.2.3 Modelling the components of the electrical system

The modelling of components of the electrical system takes into account the time-scale of the phenomena to be represented.

5.2.3.1 Network

The various elements of the network, such as lines, cables and transformers, are representated as for the load flow calculations.

5.2.3.2 Generating units

The representation of the generating units aims not to simulate their internal operation but to ensure correct behaviour as seen from the power system:

- For the electrical part, voltage control is assumed to be perfect, which means that the voltage at the terminals is equal to the reference value possibly corrected by the action of internal angle or excitation current limiters.
- For the mechanical part, the model must be adapted to the type of plant considered: hydroelectric, thermal fossil fuel or nuclear generating plant.

The problem here is to obtain sufficiently refined modelling whilst remaining compatible with the constraints of complexity and calculating

84 LONG-TERM DYNAMICS

time involved when we consider a power system containing a large number of machines. The approach will consist of modelling the turbine and the items upstream of the turbine by block diagrams.

Depending on the accuracy desired, we may use unsophisticated representations with only a few block diagrams or much more complex models which are obviously very different depending on the type of generating set considered: thermal, nuclear or hydroelectric. For example, for a 900 MW PWR nuclear unit, one could consider images of the internal thermodynamic variables (pressure, enthalpy) to model the dryer/superheater at the outlet from the high pressure casing, to represent the valves and gates as a function of the flow conditions of the steam passing through them, and all the regulating devices. One can thus obtain an overall model of the 30th or 40th order comprising very many nonlinearities.

In an appendix to this chapter, examples are given of models of a classic thermal generating set, a nuclear power unit and a hydroelectric generating set.

5.2.3.3 Modelling the loads

In classic load flow calculations, loads are represented by consumptions P_{co} and Q_{co} which can be considered as constant under normal conditions since the frequency f and the voltage U_c at the terminals of the loads are constant. Now, when the limit of power which can be transmitted is exceeded, there is no longer a mathematical solution. In reality, there is a physical solution for which the voltage U_c at the terminals of the load is reduced and the power consumed is lower than the power in the normal state. This is the voltage collapse phenomenon at which the on-load tap changers of transformers supplying power to the customers reach their limit and can no longer keep the voltage U_c constant.

A better representation of the loads, permitting simulation of this phenomenon, involves the use of admittances. The loads are thus modelled by constant admittances Y_c, linked to transformers whose transformation ratio ρ is changed by the action of the tap changers in such a way that U_c is close to the rated voltage U_{cn} (Figure 5.1).

The range of adjustment of the tap changers imposes the limits

$\rho_{min} \ll \rho_{max}$.

Moreover, the normal power values P_{co} and Q_{co} must be modified when the frequency varies by $\Delta\omega$ in relation to the rated frequency, to take into account the load shedding:

$$P'_{co} = P_{co}(1 + k\Delta\omega)\alpha$$

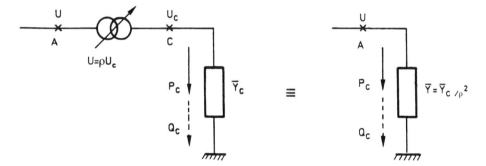

Fig. 5.1 Modelling the loads supplied by a transformer with a tap changer.

where α is the load shedding coefficient. Hence

$$Y'_{co} = \frac{(P'_{co} - jQ'_{co})}{U^2_{cn}}$$

gives the admittance corresponding to these new power values at the rated power supply voltage. $Y'_0 = Y_{co}/\rho^2$ is the equivalent impedance seen from the network.

We have:

$$Y'_0 = Y_0(1 + k\Delta\omega)$$
$$Y''_0 = (G_0 - jS_0)(1 + k\Delta\omega) \text{ with } Y_0 = G_0 - jS_0$$

The power values consumed by the node i where the voltage is U_i are written as follows:

$$P_i = G_{0i}(1 + k\Delta\omega)U_i^2$$
$$Q_i = S_{0i}(1 + k\Delta\omega)U_i^2$$

The long-term dynamic model will provide the power actually consumed, $P_i = P_{ci}$ and $Q_i = Q_{ci}$ and the voltage U_i at node i. From this we deduce the voltage $U_{ci} = U_i/\rho$ at the terminals of the load, comparison of which with U_{cn} will determine the possible change of tap resulting in a modification of ρ.

As an example, a long-term dynamic model has been used to simulate the behaviour of the load frequency control to the power system 'France' interconnected with the system 'Europe'.

86 LONG-TERM DYNAMICS

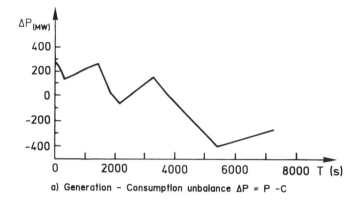

a) Generation − Consumption unbalance ΔP = P −C

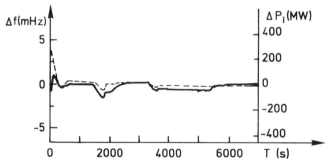

b) Frequency deviation Δf (——)
   Power flow exchange on international interconnection (− − − −)

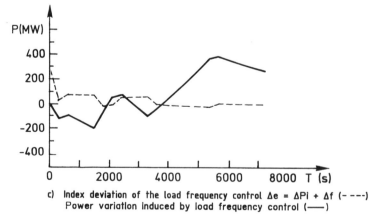

c) Index deviation of the load frequency control Δe = ΔPi + Δf (− − − −)
   Power variation induced by load frequency control (——)

Fig. 5.2 Simulation using a long-term dynamic model of a large load increase during a winter day.

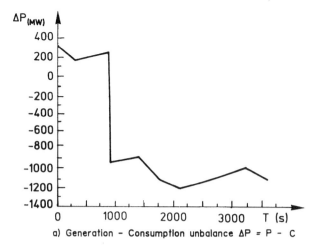
a) Generation - Consumption unbalance ΔP = P - C

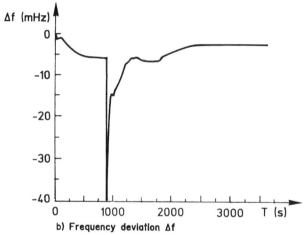
b) Frequency deviation Δf

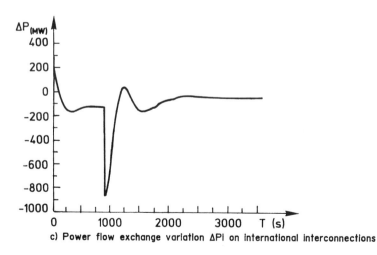
c) Power flow exchange variation ΔPi on international interconnections

Fig. 5.3 Simulation using a long-term dynamic model of the case in Figure 5.2 with tripping of a 1300 MW generating unit at time $t = 15$ min.

88 LONG-TERM DYNAMICS

Two cases have been investigated:

- a large load increase during a winter day (Figure 5.2);
- the same case with tripping of a generating unit of 1300 MW (Figure 5.3).

APPENDIX 5.A   REPRESENTATION OF A CLASSIC THERMAL UNIT WITH A DRUM BOILER

The model adopted (Figure 5.A.1) takes into account the fuel supply and the drum boiler in a very simple way.

Figure 5.A.2 also shows the speed (and power) control part, the loops for positioning the valves, and a simple model of the different turbine casings.

The main physical variables are:

Po	power reference
Pe	electrical power delivered by the generating set
Pm	mechanical power
DV	steam flow at the inlet to the turbine
R	HP opening
RMP	MP opening
$\Delta\omega$	50 Hz % speed difference (governor)
P_{To}	turbine admission reference pressure
P_T	turbine admission pressure
P_b	drum pressure
N	power frequency control level

where LP, MP and HP refer respectively to low, medium and high pressure.

The main data required for this model are then (the orders of magnitude of these parameters are indicated in brackets):

T_3	crossover time constant ($\simeq$1 s)
T_2	reheater time constant ($\simeq$1 s)
T_1	admission time constant ($\simeq$0.5 s)
T	servomotor time constant (<0.1 s)
C	capacity of the spherical tank (200–400 s)
T_f	boiler equivalent time constant (combustion, thermal exchanges) ($\simeq$30 s)
D	preparation time: for fuel oil ($\simeq$6 s) for coal ($\simeq$20–60 s depending on the type of combustion)

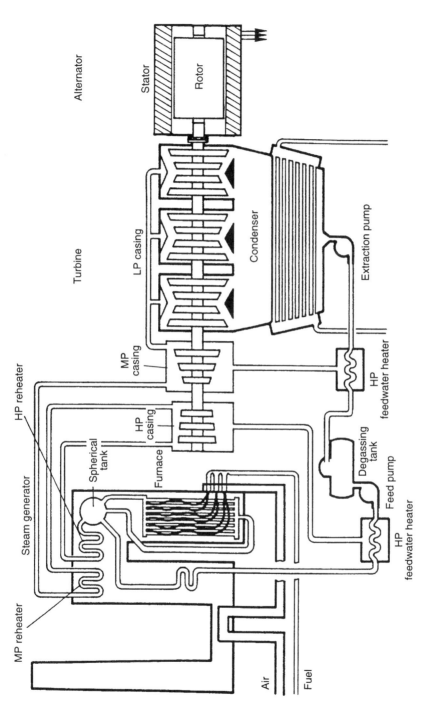

Fig. 5.A.1 Diagram showing the principles of a classic thermal unit with drum boiler.

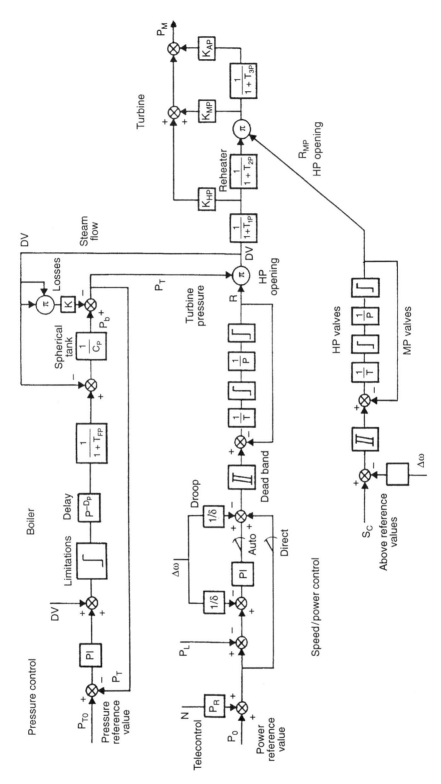

Fig. 5.A.2 Classic thermal generating unit: modelling items upstream of the turbine.

S	HP droop (4%)
S_2	MP droop (1%)
K_{HP}, K_{MP}, K_{BP}	contribution of each casing (e.g. 0.3–0.4–0.3)
K	coefficient of superheater losses ($\approx 0.08\,s$)
N	level of secondary power frequency control
P_r	participation of the generating set in telecontrol ($\pm 10\%$)

It can be noted from these data that certain variables can have widely varying values depending on the units. In particular, this is the case for the delay corresponding to consideration of the fuel preparation sytem, for which there are several alternatives, depending on the type of fuel.

In the special case of coal, several operations are performed, sometimes in different ways:

- conveying the raw coal from the silos to the crushers: the conveying speed can permit adjustment of the volume flow;
- drying the coal;
- crushing.

After crushing, the coal can be directly sent to the burners (direct combustion) or stored temporarily in silos (indirect combustion). In general this last solution is not used much.

The pulverised coal can be conveyed with the primary air or partly with recycled combustion gas. The flow of pulverised coal can also be adjusted as a function of the operating conditions.

APPENDIX 5.B REPRESENTATION OF A PRESSURIZED WATER REACTOR (PWR) NUCLEAR UNIT

The model proposed (Figures 5.B.1 and 5.B.2) is directly inspired by a linearized version of a pressurized water reactor with one single secondary loop.

The diagram shows:

- the turbine and speed and power control part (similar to the classic thermal version);
- the steam generator modelled using a time constant T_a (close to 25 s), corresponding to the volume of steam in the secondary of the steam generator (GV). It may be noted that this volume is much smaller than

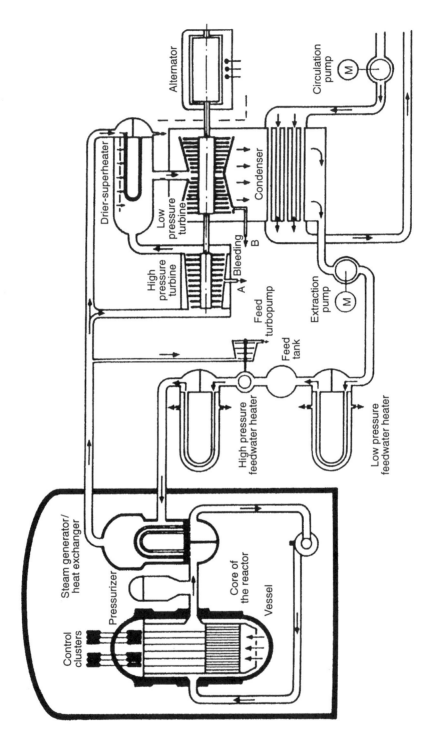

Fig. 5.B.1 Diagram of a nuclear power plant with pressurized water reactor.

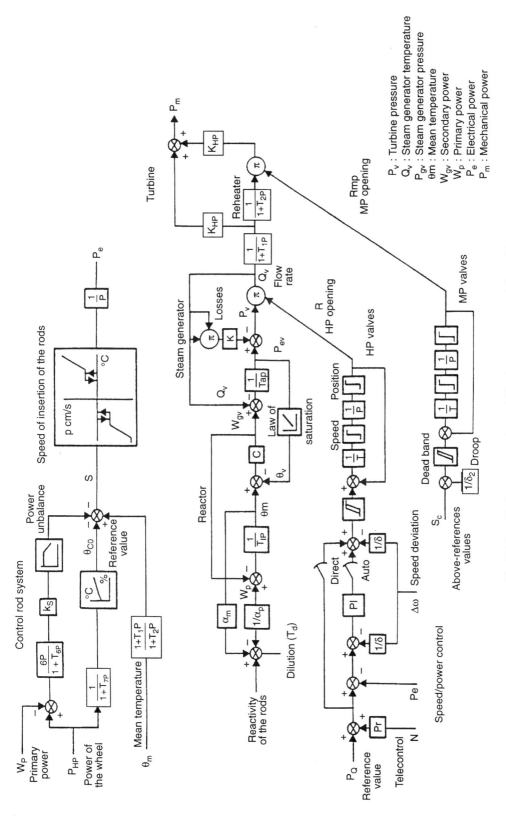

Fig. 5.B.2 Nuclear power unit: modelling items upstream of the turbine.

that contained in the spherical tank of a classic thermal boiler. One can also note the balance of power transmitted to the secondary circuit (W_{gv}) and the power delivered by the core (W_p);

- the reactor: the difference between the power transmitted to the secondary of the steam generator and the power supplied by the core affects the mean core temperature (θ_m), the value of the integration time constant T_i being of the order of 25 s.

The reactivity balance (influencing the level of power delivered by the reactor W_p) takes into account the effect of temperature, the position of the control rods and possibly the dilution of the boron.

A simple model of the system activating the control rod of the 'classic' type is also indicated: the variables serving to control the movement of the rods here are the core temperature, the primary power W_p delivered by the reactor, the pressure of the first HP wheel, which gives an image of the mechanical power at the turbine shaft; the difference $W_p - W_{mech}$ indicates the imbalance between the thermal power generated and mechanical power consumed and serves to coordinate the operation of the reactor and the turbine.

It should be emphasized that, depending on the state of progress of the fuel cycle, some parameters vary to an extent which cannot be ignored: α_p from $-0.22\%\,°C^{-1}$ to $-5\%\,°C^{-1}$ and α_m from $0\%\,°C^{-1}$ to $80\%\,°C^{-1}$ for example.

Such a model operates in variable mode, meaning that all the initial variables (in particular after integrators) must be calculated step by step, at the time $t = 0$ (equilibrium) of the simulation (working backwards from the power generated). The model then uses the deviations in relative to these initial values.

APPENDIX 5.C   MODELLING A HYDROELECTRIC POWER UNIT

Here too a distinction can be made between the portion governing the speed and controlling the position of the valves and the part upstream of the turbine (deemed to take into account hydraulic phenomena) (Figure 5.C.1).

For the speed governing–valve positioning part, a fairly simple arrangement was used:

- for speed governing, the difference in speed and its derivative are taken into account (accelerometric term m);

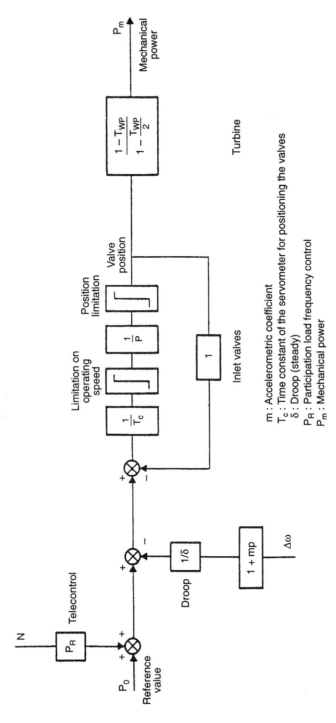

Fig. 5.C.1 Hydroelectric power unit: modelling equipment upstream of the turbine.

- for the positioning of the regulating device, a single positioning loop was taken (as for Francis turbines): other machines however have multiple positioning loops;
- only steady droop is considered (in some countries temporary droop is also found, achieved by adapting the positioning loop by a transfer function in addition to the single feedback of Figure 5.C.1).

The hydraulic part is represented using a linear model (transfer function with non-minimal phase shift). T_w varies depending on the effective head and is often of the order of 1–3 seconds (see below). Some authors advise correcting this transfer function according to the operating point and using for example a function of the type:

$$F_{(p)} = \frac{A_0 - 1.5T_{wp}}{1 + T_{wp}}$$

with A_0 variable, taking into account the nonlinearities of the mechanical power curve as a function of the opening of the valves.

Calculation of T_w (water constant)

The water constant T_w is associated with the acceleration time of the water in the penstock between the turbine intake and the reservoir. The basic relationship for this water constant is:

$$T_W = \frac{LV}{H_T g}$$

where:

    L = length of the penstock (m)
    V = speed of the water (ms^{-1})
    H_T = total head (m)
    g = constant of the acceleration of gravity (ms^{-2})

It is practical to eliminate the speed of the water and express T_w as a function of the power generated:

$$P = \rho V A H_T e$$

where:

ρ = relative density of the water
A = cross-section of the penstock
e = efficiency (turbine + alternator)

$$T_W = \frac{PL}{AH_T^2 \rho g}$$

T_w is generally between 1 s and 3 s.

FURTHER READING

Astic J.Y., Bihain A. and Jerosolimski M. (1994). The mixed Adams–BDF variable step size algorithm to simulate transient and long-term phenomena in power systems. *IEEE SM 93, IEEE Trans. Power Systems*, **9**(2), May.

Barret J.P., Pioger G. and Testud G. (1976). Simulations par programme de calcul numérique du comportement dynamique d'un réseau sur des temps longs, Conférence Canadienne sur les Communications et l'Energie, Montréal, October.

Danidson D.R., Ewart D.N. and Kirchmayer L.K. (1994). Long term dynamic response of power systems. An analysis of major disturbances. *IEEE*, **PAS-94**.

EPRI, Long Term Power System Dynamics, EPRI Research Project 90.7, Vols 1 and 2.

Lamont W. and Schulz R.P. (1979). *Long Term Power System Dynamic Research*, EEI Electrical System and Equipment Committee, Phoenix, February.

Pioger G. and Testud G. (1977). Long term dynamic behaviour of the network, Symposium on Computer Applications in Large Scale Power Systems, Calcutta.

Stubbe M., Bihain, A., Deuse J. and Baader J.G. (1989). STAG, a new unified software program for the study of the dynamic behaviour of electrical power systems. *IEEE Trans. Power Systems*, **4**(1), February.

Vernotte J.F., Panciatici P., Meyer B. *et al.* (1995). High fidelity simulation of power system dynamics. *IEEE Computer Applications in Power*, **8**(1), January.

6

STABILITY AND ELECTROMECHANICAL OSCILLATIONS

6.1 INTRODUCTION

This chapter deals with problems for which it is necessary to represent the oscillations of the machines about their position of equilibrium corresponding to the synchronous condition. It is therefore no longer possible, as for the long-term dynamic conditions (Chapter 5), to adopt the hypothesis of linked rotors. We assume that there is still a frequency common to the entire power system but the speed of rotation of a machine is an unknown quantity specific to itself.

Classically, two major classes of stability phenomena are distinguished:

- Transient stability: this corresponds to large-amplitude movements of machines during sudden disturbances, for example a short-circuit, causing considerable imbalance between the driving torque and the opposing torque.

- Small signal stability: this corresponds to the normal operating conditions of the power system in the presence of normal low-amplitude fluctuations of electrical or mechanical variables.

The system is stable if it retains or resumes a position of equilibrium where all the machines are in synchronism.

The instability of a machine can be manifested in two forms corresponding to two different causes:

TRANSIENT STABILITY 99

- Monotonic instability: the machine loses stability by monotonic deviation from its position of equilibrium. This type of instability is caused by insufficient synchronising torque.
- Oscillatory instability: the machine loses stability by oscillating about its position of equilibrium. This type of instability is caused through a lack of damping.

The two classes of stability, transient and small signal stability, obey the same laws of physics. They differ in the following aspects:

- from the viewpoint of operation of the power system:
  - transient stability corresponds to a given disturbance, accompanied by the risk associated with its degree of severity and the probability of its appearance: this is reflected in the definition of the most severe disturbance we wish to tolerate;
  - small signal stability, correspond to the normal conditions of operation, must necessarily be assured.
- from the viewpoint of simulating the phenomena:
  - transient stability relating to large movements of machines means it is necessary to consider thresholds and nonlinearities;
  - small signal stability allows linearization of the equations around the operating point.

The range of frequencies involved extends from a few tenths of hertz (0.2–0.3) to a few hertz (5–6).

In this frequency band, it is the machines which determine the dynamic behaviour of the system, so they must be represented in detail.

Figures 6.1 and 6.2 show examples corresponding to these two types of stability.

After the tripping of two outgoing lines, the power produced by the generating sets A can no longer be transmitted onto the system. The instability is manifested by divergent oscillation of the position of the rotors of the sets A in relation to the rest of the system.

6.2 TRANSIENT STABILITY

This concerns phenomena with a time constant between approximately 0.1 s and 10 s. The simulation of these will consider as instantaneous the high-speed phenomena (time constant <0.10 s) such as transients inherent in the power system. Slow variables (time constant >10 s), such as certain parameters of steam generators, will be considered as constant.

100 STABILITY AND ELECTROMECHANICAL OSCILLATIONS

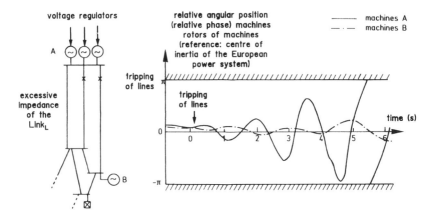

Fig. 6.1   Illustration of static instability.

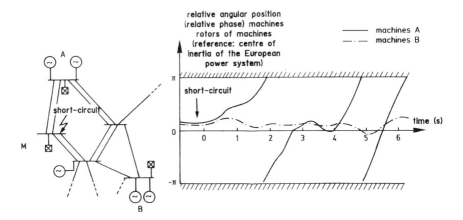

Fig. 6.2   Loss of synchronism following a busbar fault cleared rather late.

6.2.1   Modelling the components of the electrical system

6.2.1.1   Representation of generating sets

These comprise an engine and an alternator (Figure 6.3).

a) Modelling the engine

Models of engines are complex and depend on the type of engine: steam turbine, gas turbine, water turbine, diesel engine, wind turbine, etc.

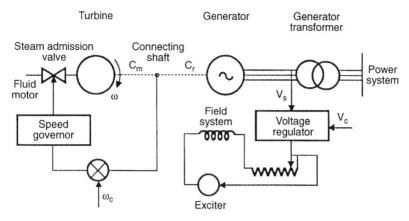

Fig. 6.3 A generating set. (C_r = opposing torque; C_m = driving torque; ω = speed of rotation; ω_c = reference speed; V_s = static voltage; V_c = reference voltage.)

Let us take for example the case of a steam turbine whose driving medium is provided by a boiler.

At the first approach, one can admit that during the period of observation of the phenomena of interest to us, the steam pressure remains constant upstream of the turbine.

The driving torque is then a function of:

- the position of the valves admitting the medium into the different casings of the turbine (HP, MP, LP);
- the time taken by the steam to travel through and its pressure reduction in the different casings.

The position of the valves is also a function of the characteristics of the speed governor and the valve control devices. The response of the driving torque to variations in speed is therefore neither instantaneous nor simple, but defined by complex transfer functions dependent on the control characteristics and physical laws controlling the interactions between the different variables.

All the transfer functions obtained lead to modelling of the type shown in Figure 6.4.

Numerous models of engines are found in the literature.

In the case in which the hypothesis of constant pressure of the driving medium upstream of the turbine cannot be adopted for the accuracy desired or the phenomenon studied, it is appropriate to represent the portion 'upstream of the turbine' by more or less sophisticated models such as those

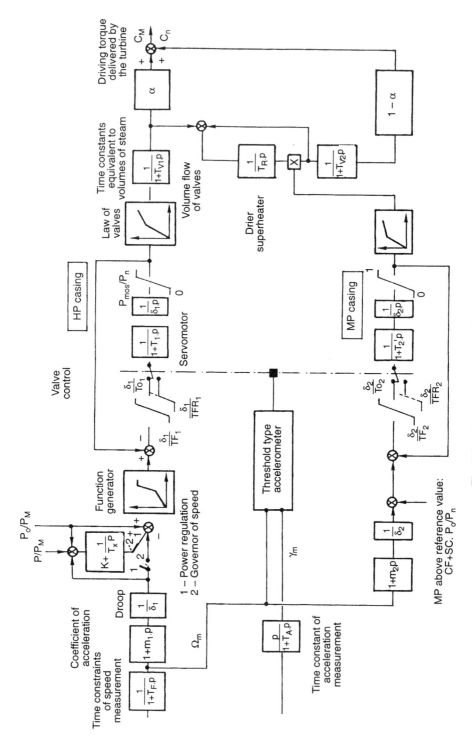

Fig. 6.4 Modelling a steam turbine assembly.

presented in Chapter 5, retaining only the elements in which the time constant has a significant influence on the results sought:

Time constant of speed measurement	T_F
Time constant of acceleration measurement	T_A
Acceleration measurement term for HP	m_1
Acceleration measurement term for MP	m_2
HP droop	δ_1
MP droop	δ_2

Power regulation:

Proportional gain	K
Integration time constant	T_x
Coefficient of the proportional term of the MP above reference value	SC
Constant term of the MP above reference value	CF

Valve control:

HP servomotor time constant	T_1
MP servomotor time constant	T_2
HP integration time constant	τ_1
MP integration time constant	τ_2
HP valve opening time	T_{01}
MP valve opening time	T_{02}
HP valve closure time	TF_1
MP valve closure time	TF_2
HP valve rapid closure time	TFR_1
MP valve rapid closure time	TFR_2

The law of valves:

HP steam time constant	T_{V_1}
MP steam time constant	T_{V_2}
Drier/superheater integration time constant	T_R
Proportion of HP torque/total torque	α

b) Modelling the generators

The choice of representation of the machines was dictated by a concern for accuracy and relative simplicity of the model. The validity of the simplifying hypotheses adopted was justified *a posteriori* by comparison of the calculation with the tests on a microsystem and the actual tests.

6. Other variables:
   - θ angle between the axis of a phase and the direct axis,
   - δ angle between the direct axis and a reference axis rotating at synchronous speed,
   - ω angular frequency,
   - ω_0 synchronous angular frequency,
   - T starting time (related to the active power)
   - C_m, C_r driving torque and opposing torque.

In the case of a machine with no damper, the electrical and magnetic equations are simplified. The equations (6.2.4), (6.2.5), (6.2.8) and (6.2.10) disappear, and in equations (6.2.6), (6.2.7) and (6.2.9) the terms i_{kd} and i_{kq} are eliminated.

To these operating equations of the machine, we must add those of the voltage regulator whose task is to keep the voltage constant at the terminals of the stator.

In the simplest case, a first-order transfer function of the form

$$\frac{\Delta v_f}{\Delta v} = \frac{G}{1 + T_e p}$$

can be admitted, G in this expression being the static gain and T_e the time constant of the assembly comprising the sensor-regulator-excitation system.

For more sophisticated regulation which can improve the stability of the generator connected to the power system, other parameters are considered, for example the variation in electric power (active power deviation stabilizer (Figure 6.6)) or the introduction of multiple additional signals, such as the electric power, the mechanical power, the speed (for example, a four feedback loop type regulator (Figure 6.7)). For severely disturbed conditions, the overexcitation and de-energization intended to limit sudden changes in voltage must also be taken into account. The transfer functions are then more complex, featuring nonlinearities such as thresholds and ceilings.

The different machines are linked together by the power system which imposes a system of linear relationships between their currents and voltages. These equations take the following form:

$$I = YV \tag{6.2.12}$$

For transient stability phenomena, the variations of impedance with frequency can be ignored, as the slip of the machines remains low.

The operating equations of the machines have been defined in relation to the electrical axes of the rotor; these are different for each machine

108 STABILITY AND ELECTROMECHANICAL OSCILLATIONS

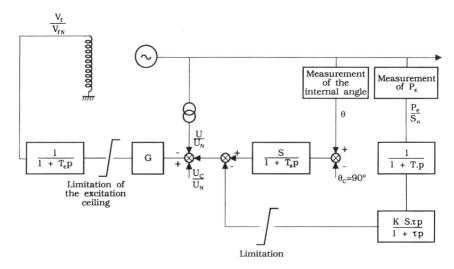

Fig. 6.6 Modelling a voltage regulator with active power deviation stabilizer.

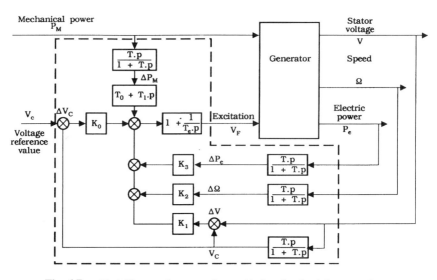

Fig. 6.7 Modelling a voltage regulator with four feedback loop regulator.

and they can also rotate at different speeds from one another during dynamic conditions. In contrast, the power system equations must be written taking common axes for the whole of the network and the machines.

Consequently, the axes of each machine are identified in relation to the reference axes situated at a certain point on the power system and rotating at the synchronous speed. A change in the coordinate axes then makes it possible to convert the different variables to apply either to the machine equations or to the power system equations.

To make the equations of the machines 'seen from the power system' easier to manage, a few simplifications will be allowed on the one hand and, on the other hand, some variables will be changed to facilitate the plotting of an operating graph of the machines under variable conditions.

Simplifications: The calculation is facilitated by two approximations in equations (6.2.1) and (6.2.2).

1. $\omega = d\theta/dt$ is replaced by ω_0 (synchronous angular frequency).

    This simplification is applicable as long as the machines do not deviate much from the synchronous speed, which is generally the case in stability investigations. Great care is required if special conditions are to be studied in which certain machines would have slip in excess of 10%.

2. The terms $d\varphi_d/dt$ and $d\varphi_q/dt$ or transformation electromotive forces are ignored in front of the terms $\omega\varphi_q$ and $\omega\varphi_d$ or rotational electromotive forces, respectively.

The expressions

$$\left(\frac{d\varphi_d}{dt} + \omega\varphi_q\right) \text{ and } \left(\frac{d\varphi_q}{dt} - \omega\varphi_d\right)$$

are the transforms in Park's transformation of winding flux derivatives of each of the three phases of the armature. The terms $\omega\varphi_q$ and $\omega\varphi_d$ correspond to the fundamental sinusoidal variation of the flux and the terms $d\varphi_d/dt$ and $d\varphi_q/dt$ which are zero under steady state conditions correspond to the aperiodic variations and harmonic sinusoidal variations of the second order. Now, the aperiodic and harmonic components die away with very low time constants and are only important for faults very close to the machines (cf. Chapter 7). The closest faults which we shall consider will be faults at the HV terminals of the generator transformers.

In view of these simplifications, the electrical and magnetic equations of the unsaturated synchronous machine without a damper can then be rewritten as follows:

$$v_d = -\omega_0\varphi_q - r_a i_d$$
$$v_q = \omega_0\varphi_d - r_a i_q$$

$$v_f = \frac{d\varphi_f}{dt} + r_f i_f$$

$$\omega_0 \varphi_d = x_d i_d + x_{ad} i_f$$

$$\omega_0 \varphi_f = x_{ad} i_d + x_f i_f$$

$$\omega_0 \varphi_q = x_q i_q$$

Changing the variables: This involves replacing the characteristic variables of the field system circuit by proportional variables:

$$e_{vf} = \frac{x_{ad}}{r_f} v_f$$

$$e_{if} = x_{ad} i_f$$

and

$$e'_q = \frac{x_{ad}}{x_f} \omega_0 \varphi_f$$

By eliminating φ_d and φ_q in the six equations of the previous paragraph, we then obtain the following four equations:

$$v_d = -x_q i_q - r_a i_d \tag{6.2.13}$$

$$v_q = e'_q + x'_d i_d - r_a i_q \tag{6.2.14}$$

$$e_{if} = e'_q - (x_d - x'_d) i_d \tag{6.2.15}$$

$$\frac{de'_q}{dt} = \frac{1}{\tau'_{do}} (e_{vf} - e_{if}) \tag{6.2.16}$$

In these expressions we have:

$$x'_d = x_a + \frac{x_{ad}(x_f - x_{ad})}{x_f} \quad : \text{ transient reactance,}$$

$$\tau'_{do} = \frac{l_f}{r_f} \quad : \text{ time contant of the field system on no-load.}$$

Equations (6.2.13), (6.2.14) and (6.2.15) can be illustrated by the vector diagram of Figure 6.8.

Under steady state conditions, we find Blondel's diagram again.

Under transient conditions, this diagram enables us to find the slowly changing variables defined above; the driving torque and speed only feature

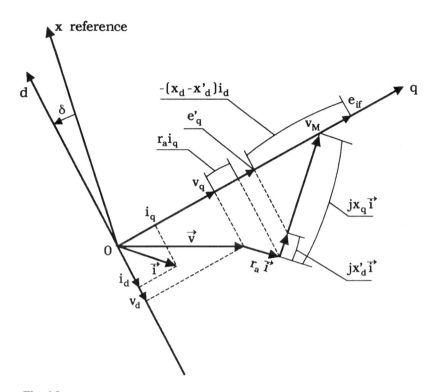

Fig. 6.8 Vector diagram of an unsaturated synchronous machine, without damper.

in the calculation of the variation of the angular position of the rotor, and the electrical angle of the rotor is represented by the position of the q axis in relation to reference axes. The field winding flux is represented by the vector e'_q. The excitation voltage, represented by e_{vf}, does not appear in the diagram. It may be noted that under steady state conditions, $e_{vf} = e_{if}$. Indeed, under steady state conditions, $d\varphi_f/dt = 0$ and $v_f = r_f i_f$.

Under transient conditions, variations of e_{vf} are determined by the action of the voltage regulator. The value of e_{vf} only features in the diagram through its action on the variation of the field winding flux, and therefore of e'_q, this action being characterized by equation (6.2.16).

6.2.1.2 Modelling the loads

Knowledge of the nature of consumptions is one of the fundamental problems when representing loads. Failing precise information, as a first approximation, the loads are considered similar to constant impedances Z

connected between the consumer nodes and earth, determined on the basis of the result of the initial distribution calculation by:

$$Z = \frac{V^2}{P - jQ}$$

In general, this representation is conservative, leading to pessimistic results.

For more realistic modelling, the variations in consumptions as a function of frequency and voltage can be considered in the following form:

$$P = K_p (V)^{\alpha_p} (f)^{\beta_p}$$
$$Q = K_q (V)^{\alpha_q} (f)^{\beta_q} \tag{6.2.17}$$

K_p, K_q are constant; α_p, α_q, β_p, β_q are coefficients of sensitivities to V and f.

For example, for a purely static load:

$$\alpha_p = \alpha_q = 2 \quad \beta_p = \beta_q = 0$$

For the phenomena of interest to us here, the influence of the frequency is still limited (Δf remains low) whilst the influence of the voltage is considerable.

At a given node, the loads are distributed over various categories characterized by equations (6.2.17).

The case of asynchronous motors

Asynchronous motors driving machines can be represented by equations (6.2.17).

Should it be necessary to have a more refined representation of the loads of the asynchronous motor, a simplified model can be used, of a single cage asynchronous motor, represented by the circuit diagram in Figure 6.9 or even more simply by the circuit diagram in Figure 6.10, by relating the magnetic leakage fluxes of the stator and rotor to the rotor only and ignoring the resistance r_1 of the stator, the value of which is still very low.

In these representations:

V_1 phase to neutral voltage applied to the stator
r_1 resistance of the stator
$l_1 \omega$ stator leakage impedance
$l_2 \omega$ rotor leakage impedance

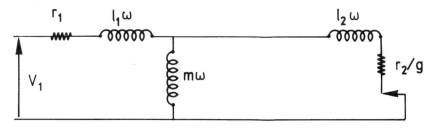

Fig. 6.9 Single-phase circuit diagram equivalent to an asynchronous motor.

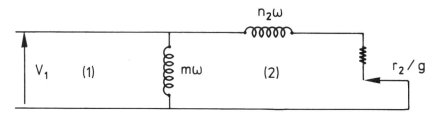

Fig. 6.10 Single-phase circuit diagram equivalent to an asynchronous motor (simplified). The stator and rotor leakage fluxes are related to the rotor (total leakage flux n_2).

$m\omega$ magnetizing impedance
r_2 resistance of the rotor circuit
g slip $= \dfrac{\Omega_n - \Omega}{\Omega_n}$
Ω speed of rotation of the motor
Ω_n speed of rotation of the rotating field of the stator (synchronous speed)
$\alpha_1 = m + l_1$ cyclic self-inductance of the stator
m mutual inductance between stator and rotor windings
$\alpha_2 = m + l_2$ cyclic self-inductance of the rotor

This representation offers the advantage of eliminating two parameters, simplifying the identification of the motor on the basis of measurements on no-load and during shutdown:

- On no-load when the opposing torque is zero, the slip is zero, $g = 0$ and the resistance r_2/g is infinite. The reactive power Q_o absorbed by the motor makes it possible to calculate the 'magnetizing' impedance $m\omega$.

- When shut down with a stalled rotor, $g = 1$, the motor absorbs the apparent power, $S = P_1 + jQ_1$, that is $S = P_1 + j(Q_1 - Q_o)$ in the branch (2), hence the calculation of $n_2\omega$ and r_2.

On the basis of the diagram of Figure 6.11, the asynchronous motors can be represented as generators without an excitation circuit, and considering the rotor as the damping circuit by using Park's equations.

In view of the isotropy of the rotor and the fact that it does not maintain a fixed position in relation to the rotating field of the stator, the same values are introduced for the corresponding variables of the direct and quadrature axes d and q.

The diagram of the asynchronous machines in Figure 6.10 can therefore be represented by the diagram of the synchronous machines in Figure 6.11 in which the impedances of the excitation circuit are infinite.

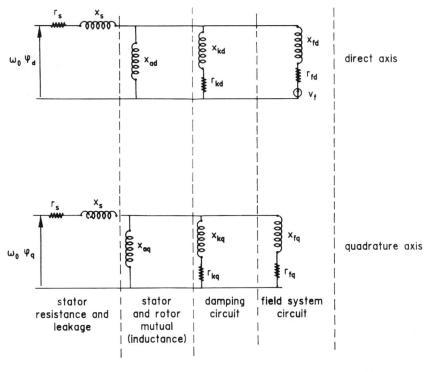

Fig. 6.11 Circuit diagram of a generator in the direct (d) and quadrature (q) axes. An aynchronous motor can be represented by such a diagram by making $X_f = \infty$ and assimilating the rotor to the damping circuit.

TRANSIENT STABILITY

The transient impedance is then equal to the synchronous impedance $x' = x_s + s_a$.

On account of their direct physical inaccessibility, the characteristics r_k and X_R of the damping circuit are determined starting from the impedance and the subtransient time constant, which can more easily be determined by measurements:

$$T'' = \frac{X_k + X_a}{\omega_{rk}} \quad \text{and} \quad X'' = X_s + \frac{X_a X_k}{X_a + X_k}$$

Note: This simplified modelling of asynchronous machines is limited to representing them as loads for low slip values. To study asynchronous machines, their starting, creeping, etc. corresponding to high slip values, it is necessary to use more precise detailed models. Moreover, the fact of ignoring the terms of $p\Phi d$ and $p\Phi q$ of Park's equations leads, in cases of severe slip, to errors as for synchronous machines.

6.2.1.3 Modelling the compensation elements

The shunt or series compensation capacitors or inductance elements are represented by their equivalent impedance.

Static VAR compensators

Static VAR compensators are represented taking into account the characteristics of their settings (Figure 6.12).

For example, let us consider a compensator comprising a variable reactance, the susceptance β_s of which can range between β_{min} and β_{max}, connected in parallel with a capacitor of fixed susceptance B_c (Figure 6.13).

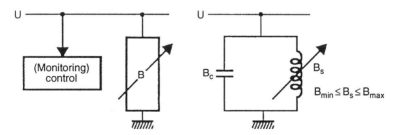

Fig. 6.12 General diagram of a static VAR compensator.

116 STABILITY AND ELECTROMECHANICAL OSCILLATIONS

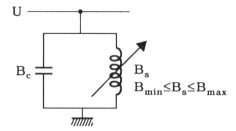

Fig. 6.13 Static compensator consisting of a variable inductance and a fixed capacitor.

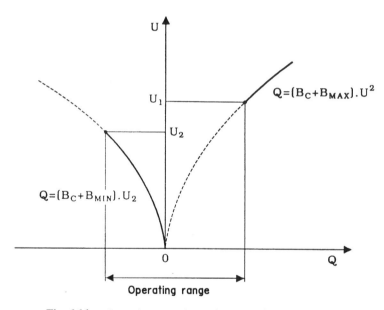

Fig. 6.14 Operating range from Figure 6.13 in the plane U, Q.

The operating characteristic of the assembly is given in Figure 6.14.

The static compensator is modelled by representing the fixed capacitor as a classic shunt compensator, the variable inductance being represented by the diagram of Figure 6.15.

In the case of a regulator of the proportional integral type, we have the circuit diagram of Figure 6.16. In this figure:

T_m time constant of the current and voltage measurement
X_{SL} droop
T_{th} time constant modelling the thyristor response

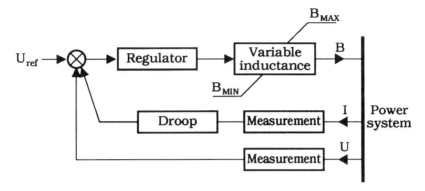

Fig. 6.15 Representation of the variable inductance of the static compensator of Figure 6.13.

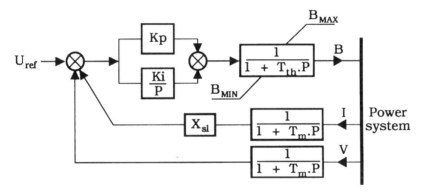

Fig. 6.16 Modelling the variable inductance of the static compensator of Figure 6.13 in the case of a proportional-integral control.

K_p proportional gain
K_I integral gain

In practice, static compensation equipments are complex, particularly their regulating devices. In conjunction with the manufacturer, modelling will be established to reproduce the dynamic phenomena with the desired degree of accuracy.

Synchronous compensators

These are represented as synchronous machines supplying zero active power and reactive power controlled by the voltage regulator.

6.2.1.4 Modelling the power system

The lines, cables and transformers are represented by elements with quadrupole localized constants of the π or T type identical to those used in the distribution calculations.

Under these conditions, the state of the power system can be represented by equations in complex numbers of the type

$$I = Y^d V$$

This representation in complex numbers assumes that one is deviating slightly from the fundamental frequency, and variations in impedance with frequency will not be taken into account.

A method of representing faults will be found in Appendix 6.B.

6.2.2 Method of solution

At the first moment following a disturbance, e'_q is retained in magnitude and in direction for each machine and the voltages and currents are calculated so that they satisfy both the power system equations and the machine equations (6.2.14).

It is appropriate to solve the power system equations by iteration, starting not from the terminals of the machine but from imaginary voltage sources v_M:

$$v_M = e'_q - (q - x'_d)i_d$$

These electromotive forces have the advantage of being carried by the q axis associated with the angular position of the rotor.

The methods of solving this are step-by-step methods. By choosing a suitable calculation interval (a few milliseconds to a few tens of milliseconds), the different characteristic variables of the system can be divided into two categories:

- variables with slow variation: these variables are either linked to inertias: driving torque, relative speed of the rotor, mechanical control devices, or they are linked to magnetic flux passing through magnetic masses, field winding flux, excitation voltage, etc.

- variables with rapid variation: currents in the different circuits, voltage at the terminals of the machines, electric power for the machines, currents and voltages for the power system.

Variables with slow variation are calculated at each interval by integration of the differential equations characterizing their changing pattern by an explicit method of the Euler or Runge Kutta type, or by an implicit trapezoidal method.

The other variables are then calculated by an iterative method starting from the algebraic equations defining them.

The discontinuities and nonlinearities call for iterations to solve the whole system of equations.

6.3 SMALL SIGNAL STABILITY

This concerns the dynamic behaviour of the power system under normal conditions subject only to slight disturbances, variations in load or generating, or the switching of devices, etc.

Under these conditions, the changes in electrical and mechanical variables have a low amplitude and the equations can be linearized about the operating point permitting the use of general research methods for linear systems.

We shall briefly present the three most commonly used methods for electrical power systems.

6.3.1 Use of the transient stability model

This model, described in section 6.2, does not use the possibility of linearizing the equations. By making small disturbances, such as opening a line with a low load or causing a small variation in load, the damping of the oscillations can be assessed.

In general, on sufficiently powerful systems, the small signal stability is no longer a problem except under certain reduced operating conditions leading to marked weakening of the power system. This stability is therefore not evaluated as such. We rely on transient stability investigations to reveal any possible problems.

It should be noted that this method does not provide an absolute guarantee since instabilities of 'static' origin can emerge at a later stage after the period of observation.

In addition, it is not easy to use when determining the optimum values of the parameters of regulating systems such as, for example, those of voltage regulator stabilizer circuits.

Its advantage is the use of one single model to evaluate the two types of stability.

6.3.2 The transmittance method

In this method, the power system and the machines are modelled in the same way as for the transient stability model but the equations are linearized about the operating point considered.

The principle of calculation is analogous to that of harmonic analysis: for each excitation frequency considered, the transmittance between the variation of the disturbed variable and the variation of the observed variable is determined.

For the conditions examined (frequency between approximately 0 and 3 Hz), the fundamental frequency (50, 60 Hz) can be considered as a carrier frequency amplitude and phase modulated by the different frequencies excited by the disturbance.

The passive elements of the power system are represented by their impedance at the fundamental frequency. The machine equations are written in operational form.

The operating equations of the system are expressed by

$$I = YV$$

where:
- I matrix of the currents injected at the different nodes;
- V matrix of the voltages at these nodes;
- Y admittance matrix of the power system.

Under disturbed conditions, it is allowed that the electrical variables remain sinusoidal and the impedances are constant.

The relationship is retained for the variations

$$\Delta I = Y \Delta V$$

The nodes where the current corresponds only to the consumption of passive loads can be eliminated by modifying the admittance matrix and finally there remain only the relationships between the voltages and currents relative to the machines, the latter being generators or dynamic loads.

The machine equations are combined with the power system equations in the general form

$$V_M = f(I_M) + g(U_M)$$

and in this expression the index M indicates that these are machine variables and U_M is a control variable taken as the input variable of the disturbance examined (excitation voltage, voltage regulator reference voltage, frequency or reference value of the speed governor, etc.).

For the variations, the expression above is written

$$\Delta V_M = f(\Delta I_M) + g(\Delta U_M)$$

Taking the Laplace transforms of the variations:

$$\Delta V_M = f(S)\Delta I_M + g(S)\Delta U_M$$

The Laplace operator S is then replaced by $j\omega_i$, ω_i being the angular frequency of the excitation which adopts discrete values in the frequency band considered.

For each value of ω_i, we solve the system of algebraic linear equations. We thus obtain the values of the transmittances (transfer functions) for the different values of ω_i considered.

We see that the results supplied are identical to those which would have been obtained by a harmonic analysis of the physical system.

On the basis of the results obtained, if an open-loop transmittance of a regulating system has been calculated, we can apply a stability criterion like the Nyquist criterion and determine the range of closed-loop stability.

On the basis of the closed-loop transmittances, it is also possible to determine the time responses to a step or an impulse by applying an inverse Fourier transform.

The advantage of this parametric calculation, compared to direct solving of the system of complete differential equations, is that the machines, not forming part of the elements observed or disturbed, do not feature in the calculation of the complete power system except through a series of coefficients linking $V_M(j\omega)$ to $I_M(j\omega)$. The calculation of these coefficients is not very sensitive to the complexity of the system of complete differential equations. We are therefore not restricted in the choice of the machine model, which can be highly detailed.

However the limitation on the complexity of the models, linked to the knowledge of the values of the corresponding parameters, emerges again here.

Figures 6.17 to 6.20 show examples of results obtained by the transmittance method.

6.3.3 The eigenvalues method

We have seen above that for stability problems, the behaviour of the power system was represented by a system of algebraic-differential equations to be solved simultaneously.

If we consider the variations about an operating point, the system can be written in the following form:

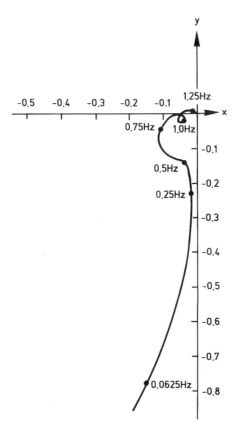

Fig. 6.17 Nyquist curve of the transmittance $(\Delta V/V_n)(\Delta V_f/V_{f_n})^{-1}$ of a 60 MW generator connected to a power system. ($\Delta V/V_n$ = variation of stator voltage with respect to rated value; $\Delta V_f/V_{f_n}$ = variation of excitation voltage with respect to rated value.)

$$\begin{pmatrix} \Delta \dot{x} \\ 0 \end{pmatrix} = \begin{pmatrix} \alpha & \beta \\ \gamma & \delta \end{pmatrix} \cdot \begin{pmatrix} \Delta x \\ \Delta y \end{pmatrix}$$

If the algebraic variables are eliminated, the matrix of state A of the system is obtained:

$$\Delta \dot{x} = (\alpha - \beta \delta^{-1} \gamma) \Delta x = A \Delta x$$

The eigenvalues of the matrix of coefficients A make it possible to determine the stability of the system. The modes of response of the system correspond to these values.

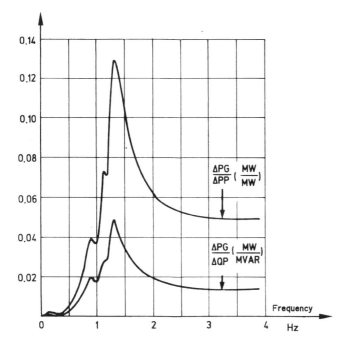

Fig. 6.18 Moduli of the transmittances $\Delta P_G/\Delta P_P$ and $\Delta P_G/\Delta Q_P$ as a function of the frequency of a generating set G close to a pulsating disturbed load P. There is an amplification factor at the resonant frequency of the generating set G (1.3 Hz) relative to the other frequency values. (ΔP_G = variation of active power of generating set G subject to the disturbance; ΔP_p, ΔQ_p = variations of active and reactive power of the disturbing load P.)

The values can be real or complex:

- a real value relates to a mode with exponential variation;
- the complex values which appear in the form of conjugate values correspond to modes of oscillation.

The imaginary part of each pair of complex roots gives the angular frequency of the oscillations.

The real part characterizes the damping of the oscillations:

- damped mode of oscillation if it is negative;
- divergent mode of oscillation if it is positive.

The matrix of state A of a system is of the size $N \times N$, N being the product of the number of generators multiplied by the number of state

124 STABILITY AND ELECTROMECHANICAL OSCILLATIONS

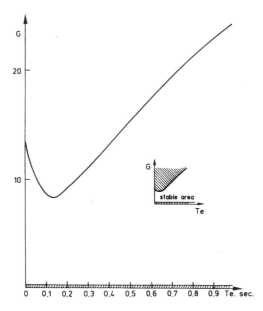

Fig. 6.19 Time response of the voltage at the terminals of a 660 MW generator connected to a system after a step of 1% of the rated value on the reference value of the voltage regulator. ($\Delta U/U_n$ = variation of stator voltage with respect to rated value.)

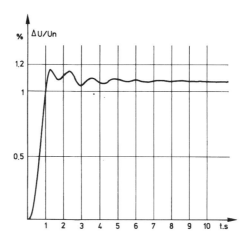

Fig. 6.20 Area of stability of a 660 MW generator connected to a power system: the transmittance of the excitation system is $G(1 + JT\omega)^{-1}$.

variables per generator. It rapidly becomes very large when the size of the electrical system increases.

The whole of the eigenvalues can be calculated at the cost of considerable constraints on the calculation, the matrix of state A not having any particular property which could be utilized.

If there are many modes of oscillation, only a few of them are slightly damped or oscillatory, and considered as unstable for the operation of the system.

Methods have therefore been developed to reduce the model studied whilst retaining the system state variables which allow us to determine the insufficiently stable modes.

In spite of everything, for problems concerning large electrical systems, such as inter-regional oscillations on interconnected power systems, we are led to use dynamic equivalents, that is groups of generators presenting coherent dynamic behaviour.

APPENDIX 6.A REPRESENTATION OF THE SATURATION

It seems very difficult to represent the saturation of the magnetic circuit exactly. However, attempts have been made to adopt the closest possible representation whilst using the same calculation procedure for machines with nonsalient poles as for salient pole machines.

It is admitted that the leakage flux of the different windings does not follow the same pattern as the air gap flux and that the latter is the only one to give rise to saturation phenomena. We consider that the magneto-motive force giving rise to the common flux is the sum of two terms: a magneto-motive force corresponding to the path of the flux in the air gap (which would be the magneto-motive force causing the flux in the absence of saturation) and a magneto-motive force which corresponds to the path of the flux in the iron which we shall call 'saturation magneto-motive force', this being considered in-phase with the air gap flux, its amplitude being a function only of the amplitude of the air gap flux and not of its phase displacement. Thus, in equations (6.2.6) to (6.2.10) relating to the flux, each component of the flux on each of the two axes is reduced by the value of the flux which, in the absence of saturation, the component of the saturation magneto-motive force would have caused in the corresponding direction. For example, equation (6.2.6) becomes:

$$\varphi_d = l_a i_d + l_{ad}(i_d + i_f + i_{kd} - i_{sd})$$

The vector diagram of the generator is then as shown in Figure 6.A.1.

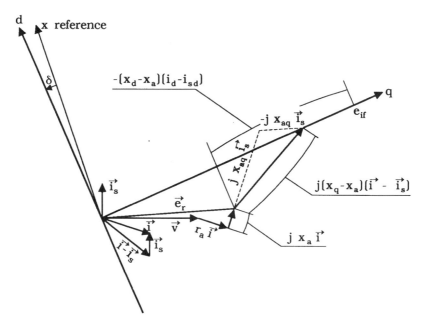

Fig. 6.A.1 Vector diagram of a saturated synchronous machine, without damper.

It may be noted that for machines with the same reactance in the two axes ($x_d = x_q$), this diagram is reduced to the Potier diagram.

In transient conditions, since the value of the air gap flux is not known when we begin calculation of the static conditions during an interval, we are led to make subiterations on the value of the saturation magneto-motive force at each stage of the iterations on the stator current value.

The saturation magneto-motive force is calculated from the air gap flux by applying a law in the form $i = k e_r^l$, the parameters k and l of which are deduced from the no-load characteristic of the generator considered.

APPENDIX 6.B REPRESENTATION OF FAULTS

The state of the power system subject to a symmetrical or asymmetrical disturbance can be represented by using a matrix Y_1^d deduced simply from the matrix Y^d defined previously in section 6.2.1.

6.B.1 Representation of the power system subject to symmetrical faults

Two examples will show how symmetrical faults can be studied:

- A three-phase short-circuit affecting node j is expressed by the condition $V_j = 0$; this means eliminating line j and column j in the matrix $\mathbf{Y^d}$.

- The three-phase tripping of a line linking the nodes j and k leads to the elimination in the matrix $\mathbf{Y^d}$ of the terms of the corresponding quadrupole.

In these two cases we are therefore led to solve a system of equations of the type:

$$\mathbf{I} = \mathbf{Y_1^d} \mathbf{V}$$

$\mathbf{Y_1^d}$ being the matrix obtained from the initial matrix $\mathbf{Y^d}$ by the operations already mentioned.

It is clear that this is generally applicable: whatever the type of symmetrical fault considered, the behavior of the disturbed power system can be investigated by a system of equations of the type:

$$\mathbf{I} = \mathbf{Y_1^d} \mathbf{V}$$

6.B.2 Representation of the power system subject to asymmetrical faults

The method used to study this type of fault calls upon the classic theory of symmetrical components [1][2] in which the electrical state of the power system is determined by using three imaginary systems whose characteristics are those of the actual power system in the positive, negative and zero-phase sequence systems.

The method of 'connections' shows that, for the type of fault considered, it is possible to take the presence of the fault into account by suitably linking the three systems: positive, negative and zero-sequence, at the location where it appears.

With regard to generators, only the positive and negative conditions are considered. The generator transformers have their upstream windings delta-connected, which prevents the passage of zero currents. The behaviour of the machines under purely negative conditions can be characterized by two phenomena: a linear relationship between the voltage and current at the terminals of the machine on the one hand, and a dissipation of energy in Joule effect losses on the stator and rotor on the other hand. The relationship between the negative voltage and negative current is established by representing the machine in the negative system by a passive impedance.

The machines being represented in the negative system by simple impedances, and not featuring in the zero-phase sequence system, the corresponding power systems are purely passive, and therefore a set of impedances reducible to a simple quadrupole will be connected to the two terminals of the fault in the imaginary positive power system. The positive voltages and currents are then calculated in the same way as for symmetrical faults, the matrix Y_1^d which represents the system with a fault in the positive system being deduced this time from the matrix Y^d of the same undisturbed power system by adding terms corresponding to the quadrupole representing the fault.

Having obtained the positive voltages at the terminals of the fault, the equations characterizing this fault make it possible to calculate the negative and zero voltages at its terminals. From this we then deduce the values of the negative and zero currents and voltages at all points on the power system using matrix equations of the corresponding systems. The voltages and currents at all the points of the power system can then be determined by superimposing their symmetrical components.

The type of fault and the negative and zero-sequence characteristics of the system are implicitly taken into account when adding the quadrupole defined above to the positive matrix of the power system.

Note

Whether the fault is symmetrical or asymmetrical, the behaviour of the system can be studied by the series of matrix equations:

$$I = Y_1^d V$$

in which the matrix of the injectors I has all its terms zero with the exception of those corresponding to the nodes where the machines are connected.

This property of the matrix I makes it possible to reduce the system of equations to be solved to a simpler form.

The system $I = Y_1^d V$ can in fact be written as follows:

$$\begin{bmatrix} I_M \\ 0 \end{bmatrix} = \begin{bmatrix} Y_{MM}^d & Y_{MR}^d \\ Y_{RM}^d & Y_{RR}^d \end{bmatrix} \times \begin{bmatrix} V_M \\ V_R \end{bmatrix}$$

in which the indices M identify the nodes where the machines are connected and R the nodes of the power system.

From this we deduce:

$$V_R = -\left[Y_{RR}^d\right]^{-1} Y_{RM}^d V_M$$

and

$$I_M = \left\{Y_{MM}^d - Y_{MR}^d \left[Y_{RR}^e\right]^{-1} Y_{RM}^d\right\} \times V_M$$

which is in the form

$$I_M = Y_2^d V_M$$

in which Y_2^d is a matrix of an order equal to the number of machines on the network: this is the matrix of the equivalent multipole of the system seen from the machines.

It is sufficient to form this at the beginning of each disturbance and the calculation at each iteration is reduced to a simple matrix multiplication (we remember that the purpose of these iterations is to make the machine variables and the power system variables compatible at every step of the calculation).

This very marked simplification of the calculations is made possible by the fact that the loads are represented by simple impedances. Any other representation of the loads would mean the complete system $I = Y_1^d V$ would have to be solved at each iteration and would also introduce extra subiterations to adapt the loads to the law of variation adopted.

As a guide, for a power system containing 12 machines and 35 network nodes, the calculation time is more or less multiplied by ten when, instead of using the system $I_M = Y_2^d V_M$, the system $I = Y_1^d V$ is solved at each iteration.

APPENDIX 6.C THE REPRESENTATION OF ROTATING MACHINES TAKING THE DAMPERS INTO ACCOUNT

To simplify the presentation, the method of calculation is explained in the case in which saturation is not considered. The electrical and magnetic equations to be taken into account are (6.2.1) to (6.2.10) for which the same approximations are allowed as in the case without a damper, that is in equations (6.2.10) and (6.2.2) $d\theta/dt$ is replaced by ω_0 and the terms with $d\varphi/dt$ are ignored. These equations then become:

$$v_d = -\omega_0 \varphi_q - r_a i_d$$
$$v_q = \omega_0 \varphi_q - r_a i_q$$

The principle of calculation is the same as that explained previously: we consider that certain variables can change rapidly and others do not. The direct axis damping circuit winding flux φ_{kd} and the quadrature axis damping circuit winding flux φ_{kq} must be added to the slowly changing variables defined previously; it is admitted that the damping circuits are equivalent to two magnetically independent circuits (one of the direct axis and the other of the quadrature axis).

The current in the direct axis damping equivalent winding i_{kd} and the current in the quadrature axis damping equivalent winding i_{kq} must be added to the rapidly changing variables. We then produce a change in the variables so that in the equations we only have variables which can be calculated from the power system. We take:

$$e'_q = \frac{x''_d - x_a}{(x_f - x_{ad})} \omega_0 \varphi_f$$

$$e'_{kq} = \frac{x''_d - x_a}{(x_{kd} - x_{ad})} \omega_0 \varphi_{kd}$$

$$e''_q = e'_q + e'_{kq}$$

$$e''_d = -\frac{x''_q - x_a}{(x_{kq} - x_{ad})} \omega_0 \varphi_{kq}$$

The variables thus defined are similar to electromotive forces. They can be considered as maintained at the first instant of a disturbance, since they are proportional to winding fluxes relating to circuits closed on a zero impedance.

We also take:

$$e_{jkd} = x_{ad} i_{kd}$$

$$e_{jkq} = x_{aq} i_{kq}$$

Equations (6.2.1) to (6.2.10) can then be reduced to the following:

$$v_d = e''_d - x''_q i_q - r_a i_d \tag{6.C.1}$$

$$v_q = e''_q + x''_d i_d - r_a i_q \tag{6.C.2}$$

$$e_{if} = \frac{x_d - x_a}{x_d'' - x_a} e_q' - \frac{x_d - x_d'}{x_d' - x_a}\left[e_q'' + \left(x_d'' - x_a\right)i_d\right] \tag{6.C.3}$$

$$e_{ikd} = -e_{if} + e_q'' - \left(x_d - x_d''\right)i_d \tag{6.C.4}$$

$$e_{ikq} = -v_d + r_a i_d - x_q i_q \tag{6.C.5}$$

$$\frac{de_q'}{dt} = \frac{x_d'' - x_a}{x_d' - x_a} \frac{1}{\tau_{do}'}\left(e_{ef} - e_{if}\right) \tag{6.C.6}$$

$$\frac{de_q''}{dt} = \frac{de_q'}{dt} + \frac{de_{kq}'}{dt}$$

$$= \frac{de_q'}{dt} - \frac{x_d' - x_a}{x_d - x_a} \frac{1}{\tau_{do}''} e_{ikd} \tag{6.C.7}$$

$$\frac{de_d''}{dt} = \frac{1}{\tau_{qo}''} e_{ikq} \tag{6.C.8}$$

In these expressions:

$$x_d'' = x_a + \frac{x_{ad}\left(x_f - x_{ad}\right)\left(x_{kd} - x_{ad}\right)}{x_{kd} x_f - x_{ad}^2} : \text{ direct axis subtransient reactance,}$$

$$x_q'' = x_a + \frac{x_{aq}\left(x_{kq} - x_{aq}\right)}{x_{kq}} : \text{ quadrature axis subtransient reactance,}$$

$$\tau_{do}'' = \frac{1}{r_{kd}} \frac{l_f l_{kd} - l_{ad}^2}{l_f}$$

$$\tau_{qo}'' = \frac{l_{kq}}{r_{kq}}$$

- e_r: electromotive force after the stator leakage impedance, corresponding to the air gap flux,
- i_s: imaginary current corresponding to the additional ampere-turns required by the saturation phenomena, assumed to be in phase with the air gap flux,
- i_{sd}, i_{mj}: components of i_s on the direct axis and the quadrature axis.

Equations (6.C.1), (6.C.2), (6.C.4) and (6.C.5) can be expressed in the vector diagram of Figure 6.C.1. The retention of the damping and field system winding fluxes is expressed by retaining an electromotive force e'' with components e_d'' and e_q''. At the beginning of each step in the calculation the magnitude and phase of this vector are known and the stator current must satisfy on the one hand (6.C.1) and (6.C.2), and on the other

132 STABILITY AND ELECTROMECHANICAL OSCILLATIONS

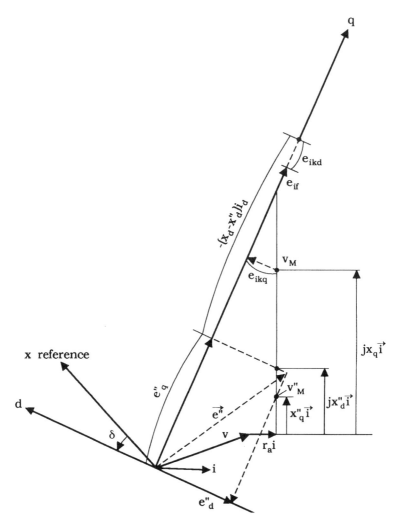

Fig. 6.C.1 Area of stability of a rotating machine, with damping.

hand the power system equations. The mechanism of the calculation can be similar to that described when there is no damping circuit, but to solve the system, at each step in the calculation, the machines are considered as imaginary sources represented in vector form by v''_M, and no longer by v_M, that is sources whose electromotive force is deduced from the voltage at the terminals by vector addition of the voltage drop due to the passing of the stator current in the quadrature axis subtransient impedance of the machine.

The advantage of this choice is that solving the power system on the basis of these electromotive forces, representing the machines by their quadrature axis subtransient reactance x_q'' directly gives the components e_d'' sought and the iterations therefore relate only to the value of e_q''. The component of v_M'' on the q axis must have a value such that:

$$v_{Mq}'' = e_q'' + (x_d'' - x_q'')i_d \tag{6.C.9}$$

Remarks

1. e_q' does not appear on the diagram.

    Under steady state conditions, the diagram of Figure 6.C.1 is reduced to the Blondel diagram on which certain specific electromagnetic forces have been identified. Indeed, the currents in the damping circuits are zero: $e_{ikd} = e_{ikq} = 0$. It is this which makes it possible, starting with the power system, to determine the initial values of the different machine variables and in particular e_{if}, hence e_q' by application of the equation (6.C.3).

    Under transient conditions, e_q' is deduced from its value at the preceding step and from the value of its derivative.

2. When the forward axis and quadrature axis subtransient reactances of the machines are equal, the iterations are eliminated and calculation is much faster. It should however be noted that the variables e_d'' and e_q'' vary from one step to another with different time constants and in particular e_q'' alone is influenced by the voltage regulation, through the intermediary of e_q.

3. When there is no damping circuit, $x_d'' = x_d'$, $x_q'' = x_q$, which leads to $e_q'' = e_q'$, $e_d'' = 0$, $e_{ikd} = e_{ikq} = 0$. We then come back to the calculation described in section 6.2.2.

REFERENCES

1. CIGRE (1966). Paper 334a, *Definition of general terms relating to stability of interconnected synchronous machines.*
2. Coneardia C. (1951). *Synchronous Machines: Theory and Performances*, Wiley and Sons, New York.

FURTHER READING

Arcidiacono V., Ferrari E., Marconato R. and Dos Ghali J. (1980). Evaluation and improvement of electromechanical oscillation damping by means of eigenvalue-eigenvector analysis–practical results in the Central Peru power system. *IEEE*, **PAS 99**.

Arnoldi W.E. (1951). The principle of minimized iterations in the solution of the matrix eigenvalue problem. *Quart. Appl. Math.*, **9**.

Arnoldi C.P. and Pacheco E.J.P. (1979). Modelling induction motor start up in a multimachine transient stability program, *IEEE PES Summer Meeting*, Vancouver, Canada.

Barbier C. and Logeay Y. (1973). Identification des réseaux de transport et des unités de production (Identification of transmission systems and generating units), *Revue Générale d'Eléctricité*, September.

Barbier C., Barret Ph., Pioger G. and Sapet P. (1968). Utilisation des calculatrices numériques pour l'étude de la stabilité des réseaux de transport d'énergie (The use of digital computers for research into the stability of power transmission systems), *Revue Générale d'Eléctricité*, July/August.

Barbier C., Carpentier L. and Saccomano F. (1978). Tentative classification and terminologies relating to stability problems of power systems. *ELECTRA* No. 56.

Barret Ph. (1972). Détermination des paramètres des machines synchrones par la méthode d'analyse fréquentielle (Determination of the parameters of synchronous machines by the frequency analysis method), *Revue Générale d'Eléctricité*, December.

Barret Ph. (1981). *Régime Transitoires des Machines Tournantes Électriques* (Transient conditions of rotating electric machines), Eyrolles, Paris.

Bauer D.L. *et al.* (1975). Simulation of low frequency undamped oscillations in large power systems. *IEEE*, **PAS 94**, March/April.

Breteton D.S., Lewis D.G. and Young C.C. (1957). Representation of induction motor loads during power system stability studies. *AIEE*, **PAS 76**.

Concardia C. and Ihara S. (1982). Load representation in power system stability studies. *IEEE*, **PAS 101**, April.

Crary S.B. (1945). *Power System Stability*, Wiley and Sons, New York.

Crary S.B., Herlitz I. and Favez B. (1958). Report on the work of Study Committee 13: System Stability and Voltage, Power and Frequency Control, CIGRE Report 347.

Cullum J. and Willoughby R.A. (1986). A practical procedure for computing eigenvalues of large sparse non-symmetric matrices, in *Large Scale Eigenvalue Problems*, eds J. Cullum and R.A. Willoughby, Elsevier Science Publishers B.V.

Dandeno P.L. (1982). General overview of steady state (small signal) stability in bulk electricity systems. *Electrical Power Energy Syst.*, **4**, 253.

Dandeno P.L. and Kundur P. (1973). A non-iterative transient stability program including the effects of variable load-voltage characteristics. *IEEE*, **PAS 92**.

Dandeno P.L. *et al.* (1973). Effects of synchronous machine modelling in large scale system studies. *IEEE*, **PAS 92**.

El Abiad A.H. and Nagappan K. (1966). Transient stability of multimachine power systems. *IEEE*, **PAS**.

Gaugeuil J.C., Caustère A., Barbier C. and Pioger G. (1971). Point de quelques questions concernant la stabilité dynamique des réseaux (A look at a few questions concerning the dynamic stability of power systems), *Revue Générale d'Eléctricité*, May.

Hammons T.J. and Winning D.J. (1971). Comparison of synchronous machine models in the study of the transient behaviour of electrical power systems. *Proc. IEEE*, **118**.

IEEE Committee Report (1968). Computer representations of exciter systems. *IEEE*, **PAS 87**.

IEEE Committee Report (1973). Dynamic models for steam and hydroturbines in power system studies. *IEEE*, **PAS 92**.

Irving E., Logeay Y. and Roquefort Y.M. (1972). Power grid network identification, *PSCC*, Grenoble.

Jordan M.E. (1979). Synthesis of double cage induction motor design. *AIEE*, **PAS 78**.

Kimbark E.W. (1956). *Power System Stability*, Wiley and Sons, New York.

Kundur P. and Dandeno P.L. (1975). Practical application of eigenvalue techniques in the analysis of power system dynamic stability problems, *Proc. Fifth Power System Computation Conference*, Cambridge, England, September.

Logeay Y., Maury F. and Roquefort Y.M. (1972). Network linear model: computer program and applied example, *PSCC*, Grenoble.

Martins N. (1986). Efficient eigenvalue and frequency response methods applied to power system small-signal stability analysis. *IEEE*, **PWRS-1**.

Nelles D. (1972). Comparison of the various mathematical generator models for the calculation of transient electromechanical phenomena. *PSCC*, Grenoble.

Obata Y. Takeda S. and Suzuki H. (1981). An efficient eigenvalue estimation technique for multimachine power system dynamic stability analysis. *IEEE*, **PAS-100**, (1), January.

Olives D.W. (1966). New techniques for the calculation of dynamic response. *IEEE*, **PAS-85**.

Pagola F.L., Pérez-Arriaga I.J. and Verghese G.C. (1989). On sensitivities, residues and participations: applications to oscillatory stability analysis and control. *IEEE*, **PWRS-4**, (1), February.

Pagola F.L., Rouco L. and Pérez-Arriaga I.J. (1990). Analysis and control of small signal stability in electric power systems by selective modal analysis. Eigenanalysis and frequency domain methods for system dynamic performance, IEEE Publication No. 90TH0292-3-PWR, February.

Podmore R., Athay Th., Germond A. and Virmai S. (1988). New techniques for analysis of power system stability, *6th PSCC Proceedings*, Darmstadt.

Racz L.Z. and Bokay B. (1988). *Power System Stability*, Elsevier, Amsterdam, Oxford, New York, Tokyo.

Rouco L. and Pérez-Arriaga I.L. (1992). Multi-area analysis of small signal analysis in large electric power systems by selective modal analysis, IEEE/PES Summer Meeting, Paper No. 92 SM 601-5 PWRS, Seattle (Washington).

Ruhe A. (1984). Rational Krylov sequence methods for eigenvalue computation. *Linear Algebra and Its Applications*, **58**.

Saad Y. (1980). Variations on Arnoldi's method for computing eigenelements of large unsymmetric matrices. *Linear Algebra and Its Applications*, **34**.

Semlyen A. and Wang L. (1988). Sequential computation of the complete eigensystem for the study zone in small signal stability analysis of large power systems. *IEEE*, **PWRS-3**, (2), May.

Smed T. (1991). Feasible eigenvalue sensitivity for large power systems, IEEE/PES Winter Meeting, Paper No. 91 SM 171-0 PWRS, New York.

Uchida N. and Nagao T. (1988). A new eigen-analysis method of steady-state stability studies for large power systems: S matrix method. *IEEE*, **PWRS-3**, (2), May.

Venikov V.A. et al. (1972). New method of simulation and numerical solution of transient phenomena in electrical systems. *PSCC*, Grenoble.

Venikov V.A. et al. (1972). Analysis of steady state stability of complex power systems by frequency response analysis. *PSCC*, Grenoble.

Verghese G.C., Pérez-Arriaga I.J. and Schweppe F.C. (1982). Selective modal analysis with applications to electric power systems, Part II: The dynamic stability problem. *IEEE*, **PAS-101**, (9), September.

Wilkinson J.H. (1965). *The Algebraic Eigenvalue Problem*, Clarendon Press, Oxford.

Wong D.Y., Rogers G.J., Porretta B. and Kundur P. (1988). Eigenvalue analysis of very large power systems. *IEEE*, **PWRS-3**, (2), May.

7

ELECTROMAGNETIC TRANSIENTS

7.1 INTRODUCTION

In Chapter 6, on stability phenomena in synchronous machines, the modelling of the electrical system represented in detail the control systems with their time constants generally greater than 100 milliseconds, the thermal hydraulic and mechanical equations of the generating sets and subtransient time constants of the fluxes in the alternator.

Beyond this, from the alternator stators to the lines, transformers and power electronics equipments (in short, for everything which forms the structure of the electrical system), modelling in the form of impedances is adopted, which presupposes that the system is subject to a succession of sinusoidal quasi-stationary conditions. In the context of transient stability, where the resolution of the mechanical equation for each synchronous machine makes it possible to describe the individual changing status of each rotor, this hypothesis is no longer strictly valid, on account of the desynchronization of the machines.

The artificial definition of a 'mean frequency', the mean of the frequencies of each machine in proportion to its starting time, allows this inconsistency to be set aside and enables us to take into account a variation in the impedances of the system (chiefly the shunt impedances which represent the loads) as a function of this mean frequency. This modelling of the system remains valid as long as the variation of the mean frequency remains low in relation to the rated frequency.

Moreover, although the electromechanical transients which are manifested in the desynchronization of the rotors can be represented using this modelling, the phenomenon which initiates this rotor transient (short-circuit, incorrect coupling) is still incompletely represented as it involves a transient response of the electrical circuits, not shown by a model using impedances.

These transient conditions are likely to reveal, for example, a transient braking torque in the first moments following the appearance of a short-circuit, which can cause loss of synchronism by underfrequency on a light machine or a synchronous compensator. This transient braking torque cannot be represented in a model in which this stator and the system are represented by impedances.

To be able to study these rapid phenomena which affect the dynamic response of the system in a frequency range up to several kilohertz, there are two classes of modelling:

- representation in the form of lumped elements, which leads to a system of nonlinear differential equations;
- representation in the form of distributed elements, where the propagation of the electromagnetic wave is taken into account. This type of representation is necessary in modelling power transmission lines and cables as soon as we become interested in phenomena with frequencies greater than 100 Hz on lines over 100 km long.

First, we shall quickly review the electromagnetic phenomena appearing in the electrical system which require modelling of one of these two types. Then the modelling of the key elements in each class will be developed: the lines with distributed elements, and transformers and synchronous machines with lumped elements. The modelling of lines shows how the propagation of waves can be taken into account in a fairly simple form in order to be compatible with the formalism of the circuits; the modelling of transformers and alternators will reveal the origin of certain nonlinearities in the system.

7.2 PHYSICAL PHENOMENA CALLING FOR MODELLING OF ELECTROMAGNETIC TRANSIENTS

For high speed phenomena, current and voltage waves can no longer be considered as sine waves at 50 Hz or 60 Hz with slow amplitude modulation. As an illustration, phenomena appearing on the system which, on account of their highly nonlinear nature or their extended frequency spectrum, require detailed representation of the dynamics of the electrical system, include lightning waves which appear following a lightning strike on one or more conductors of a line, or an overhead earth wire, or the ground close to a line. The propagation of the resulting waves calls for a representation of the lines, towers, surge impedance of the earth connections derived directly from Maxwell's equations, and modelling of the nonlinear lateral losses in the dielectric (corona effect). The modelling of the air gap and the surge impedance of the tower can also be required if one wishes to simulate back flashover, which gives rise to waves of a steeper form than the lightening wave itself. To obtain an idea of the form of the lightning wave, it is sufficient to know that this is synthesized in the laboratory as a biexponential wave with a rise time of $1.2\,\mu s$ and a decay time of $50\,\mu s$.

Switching overvoltages appear during switching operations on the system, particularly the opening and closing of circuit breakers, energization of lines on no-load or of saturable transformers. When circuits are closed, a transient condition arises in the system, the approximate pattern of which, in the first few moments, ignoring the nonlinearities of the system, is the response of the system to a voltage step when closing takes place at maximum voltage, or the response to a voltage ramp when closure takes place in the vicinity of the voltage zero.

When circuits are tripped, it is discontinuities of current caused by the chopping effect of the circuit breakers, or more generally the discontinuity of the derivative di/dt during natural extinction of the current as it passes through zero, which creates a transient overvoltage which then propagates along the lines.

7.2.1 Ferroresonance

This is a nonlinear phenomenon which can present several stable operating states and pass suddenly from one of these states to another following an infinitesimal variation in one of the parameters of the system. Bifurcation theory allows this phenomenon to be explained overall on systems of small size. The conjunction of a (shunt or series) capacitance and a saturable inductance (magnetic circuit of a transformer) can show this phenomenon in its full complexity, the nonlinear differential equation representing this simplified circuit being of the Duffing equation type.

7.2.2 Subsynchronous resonance

Subsynchronous oscillations are oscillations at a frequency lower than the synchronous frequency (50 Hz or 60 Hz, depending on the system), the damping of which can in certain conditions become weak, or even negative.

They are caused by nonlinear interaction between the normal modes of mechanical torsional vibrations on the shafts of large turboalternators and the resonant frequencies of the electrical system. The nonlinearity of the interaction between the mechanical and electrical modes permits a transfer of power between the electrical mode at 50 Hz and the oscillating mechanical mode, which can lead to divergent torsional oscillations capable of damaging the shafts of the turboalternators.

This far from complete typology of the electromagnetic phenomena which may affect the electrical system reveals the need for a more complex level of modelling than representation in the form of impedances at 50 Hz or 60 Hz, which can take into account the phenomena of propagation at high frequency and the many nonlinearities present in the electrical system.

7.3 TYPES OF MODELLING USED

Here we shall describe the modelling of components of the system which have a significant effect on the phenomena described in the preceding paragraph.

The modelling of transmission lines will show how to take into account propagation on the system, taking as a basis Maxwell's equations, and simplifying and homogenizing them sufficiently to integrate them into the elements of the system as a whole (transformers, synchronous machines) in which propagation is ignored. The system is thus represented by circuits governed by nonlinear differential equations linking macroscopic variables such as the voltages (phase-to-earth or between phases) and the currents flowing into the conductors.

The modelling of transformers and synchronous machines in the form of a group of discrete magnetic and electric circuits will make it possible to explain how the nonlinearities are taken into account.

7.3.1 Modelling overhead lines

To model overhead lines, we start with Maxwell's equations and endeavour, by simplifications associated with hypotheses of geometrical symmetry and conductivity of the conductors and of the ground, to obtain relationships between macroscopic variables such as voltages and currents, which will then make it possible to model the electrical system as a whole in the form of a differential system, using Kirchhoff's laws.

It is assumed that the media (air, the ground) are linear, homogeneous and isotropic, hence

$$\begin{cases} \boldsymbol{D} = \varepsilon_0 \varepsilon_r \boldsymbol{E} \\ \boldsymbol{B} = \mu_0 \mu_r \boldsymbol{H} \end{cases} \quad (7.3.1)$$

where D is the electric induction, E is the electric field, B is the magnetic induction, H is the magnetic field, ε_0 is the dielectic constant, ε_r is the relative dielectic constant, μ_0 is the magnetic permeability, μ_r is the relative magnetic permeability and in which $\varepsilon_0 \mu_0 c^2 = 1$, and the conductors follow Ohm's law, namely

$$j = \sigma E \tag{7.3.2}$$

We then have the two Maxwell èquations linking E to B and H to D:

$$\begin{cases} \nabla \wedge E = \dfrac{-\partial B}{\partial t} \\ \nabla \wedge H = \dfrac{j + \partial D}{\partial t} \end{cases} \tag{7.3.3}$$

in which j is the current density in the conductors and in the ground.

The conservation of electric charges in a closed volume is written as follows:

$$\nabla \cdot j + \frac{\partial \rho}{\partial t} = 0 \tag{7.3.4}$$

The divergence of a curl being zero, by taking the divergence of (7.3.3) and using (7.3.4), we obtain (7.3.5), generally designated as Maxwell's third and fourth equations.

$$\begin{cases} \nabla \cdot B = 0 \\ \nabla \cdot D = \rho \end{cases} \tag{7.3.5}$$

The classic solution of (7.3.3) and (7.3.5) takes the scalar potential V and vector A:

$$\begin{cases} E = -\overline{\nabla V} - \dfrac{\partial A}{\partial t} \\ B = \nabla \wedge A \end{cases} \tag{7.3.6}$$

To deal with the problem, simplifying hypotheses will be adopted, considering:

- the geometry of the lines (hypothesis 1),
- the conductivity of the conductors and values of ε_r and μ_r (hypothesis 2),

142 ELECTROMAGNETIC TRANSIENTS

- and, above all, the frequency range dealing with the phenomena. It is therefore useful to remember that to represent the phenomena mentioned in 7.1, a frequency band of 0–10000 Hz is generally sufficient (hypothesis 3).

We shall give details of these various hypotheses taking account their importance in modelling overhead lines.

Hypothesis 1

The complex formed by the conductors of the line and the ground (assumed to be horizontal and homogeneous) forms a waveguide for plane longitudinal waves whose direction is that of the line. The transversal dimensions (mean height in relation to the ground, and distance between conductors) being low in relation to the wavelengths of the phenomena studied, propagation in a plane perpendicular to the direction of the line is ignored.

Numerical application

For a line with a maximum height of 30 m, the transverse diameter is of the order of 100 m (electrical images plus skin effect, which we will see later). For accuracy of the model of the order of ±5% (more than sufficient) and a speed of propagation of the wave

$$\frac{1}{\sqrt{\varepsilon \varepsilon_0 \mu \mu_0}} \cong c$$

a maximum frequency of 30 kHz is found for the phenomena falling within the context of these hypotheses. Beyond these frequencies, much more complex three-dimensional models must be envisaged, taking into account the discontinuities constituted by the towers, etc.

Hypothesis 2

The resistivity values per unit of length of the conductors are low:

- high conductivity for conductors of small cross-section (phase conductors or overhead earth wires),
- large cross-section for the low conductivity conductor which is the ground (σ varies from 100 to 10^{-2} S m^{-1}).

The longitudinal field

$$E_x = \frac{1}{\sigma} j_x$$

is then low in relation to the field perpendicular to the surface of the conductors and the ground, therefore, outside the conductors and the ground

$$E \cong E_{\text{electrostat}} = -\nabla V \quad \text{and} \quad U_a = \sum_b \Lambda_{ab} Q_b$$

the relationship between the surface voltages and charges, with Λ_{ab} slightly different from the electrostatic influencing coefficient.

Hypothesis 3

The electric charges are surfaces charges. From (7.3.4), (7.3.5), (7.3.1) and (7.3.2) we obtain:

$$\frac{\sigma \rho}{\varepsilon \varepsilon_0} + \frac{d\rho}{dt} = 0$$

The time constant

$$\frac{\varepsilon \varepsilon_0}{\sigma} = \frac{10^{-9}}{36\pi \times 10^{+8}} \simeq 8.8 \times 10^{-20} \text{ s}$$

for phase conductors or overhead earth wires, which justifies $\rho = 0$.

In the ground $\varepsilon < 100$, $\sigma > 10^{-2}$. The time constant there thus remains lower than

$$\frac{10^{-7}}{36\pi \times 10^{-2}} \simeq 8.8 \times 10^{-8} \text{ s}$$

which, for phenomena below 10 kHz still justifies the approximation $\rho = 0$ in the ground.

From the second equation of (7.3.3), from (7.3.1) and the continuity of the tangential component of B, in the absence of surface currents, we obtain:

$$\varepsilon_i \mu_i \left(p + \frac{\sigma_i}{\varepsilon_i \varepsilon_0} \right) E_{ni} = \varepsilon_e \mu_e p E_{ne} \tag{7.3.7}$$

in which the index i represents the medium inside the conductors (and in the ground) and the index e represents the exterior, and p the Laplace operator.

E_{ni} and E_{ne} are electric fields perpendicular to the surface of the conductors, directed towards the interior and exterior of these conductors, respectively.

From (7.3.7) we obtain:

$$\frac{E_{ni}}{E_{ne}} = \frac{\varepsilon_e \mu_e}{\varepsilon_i \mu_i + \mu_i \sigma_i / p\varepsilon_0} \qquad (7.3.8)$$

In general $\varepsilon_e = \mu_e = \mu_i = 1$. For phase conductors and over-head earth wires

$$\frac{E_{ni}}{E_{ne}} \cong \frac{1}{1 + \sigma/p\varepsilon_0} \quad \sigma \cong 10^8 \quad \varepsilon_0 = \frac{10^{-9}}{36\pi}$$

and $p \leq 2\pi \times 10^4$ if $f \leq 10\,\text{kHz}$, hence $E_{ni}/E_{ne} \cong 1.8 \times 10^{-12}$.

For the ground $\sigma_i > 10^{-2}$ and $\varepsilon_i < 10^2$, hence $E_{ni}/E_{ne} \cong 1.8 \times 10^{-6}$.

The electric field, in accordance with hypothesis 2, is therefore perpendicular to the external surface of the conductors and thus coincides there

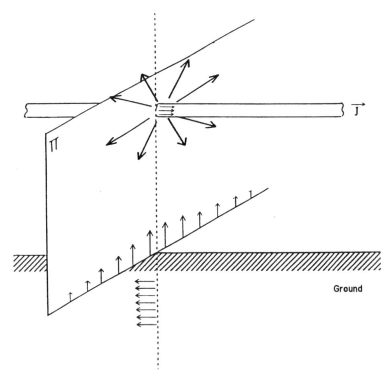

Fig. 7.1 Pattern of the field E (single-phase line).

TYPES OF MODELLING USED 145

with the electrostatic field $-\nabla V$. It is parallel to the direction of the line in the conductors and in the ground, the equipotentials then being the cross-sections of these conductors. This makes the concept of voltage between phases or phase-to-earth voltage pertinent and permits the linking of these surface voltages between conductors with the surface loads by

$$U_m = \sum_n \Delta_{mn} Q_n$$

in which the Δ_{mn} values are the coefficients of electrostatic induction between conductors, obtained by the electrical image method. Figure 7.1 gives an idea of the pattern of E in a plane orthogonal to the line. Figure 7.2 explains the electrical images and the calculation of the Δ_{ij} values.

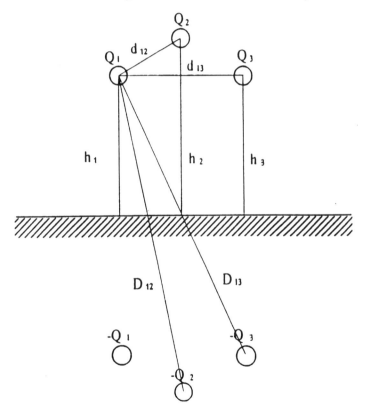

Fig. 7.2 Calculation of the Δ_{mn} values by electrical images.

$$\Delta_{ii} = \frac{1}{2\pi\varepsilon\varepsilon_0} \ln \frac{2h_i}{r_i}$$

$$\Delta_{ij} = \frac{1}{2\pi\varepsilon\varepsilon_0} \ln \frac{D_{ij}}{d_{ij}} \quad i \neq j$$

To be able to interface these line and cable propagation models with those of other elements of the network, represented by a discrete assembly of circuits, we shall establish relationships between the macroscopic variables U and I, the total current in the conductors and the ground.

According to (7.3.8), E_{ni} being low, the flow-lines of current in the conductors are parallel to the direction of the line; therefore the vector potential A is essentially parallel to this direction.

Let us take the case of a conductor above the ground. By writing (7.3.6) in cylindrical coordinates based on the axis of the conductor, in a plane perpendicular to the ground, passing through the conductor and outside it, we have:

$$B_x = 0; \quad B_r = \frac{1}{r}\frac{\partial A_x}{\partial \theta}; \quad B_\theta = -\frac{\partial A_x}{\partial r} \qquad (7.3.9)$$

$E_\theta = 0$ (equipotentials i.e. plane perpendicular to the line) and $E_r = -\partial U/\partial r$.

As A_x is symmetrical in relation to the plane perpendicular to the ground passing through the conductor, $B_r = 0$.

In the hypothesis of low damping values in which the wave is propagated as e^{-kx} and in which $k \cong p/v$, we have $J = QV$ in which J is the current in the conductor and Q the charge at its surface (taken from (7.3.4)) which becomes

$$\frac{\partial j}{\partial x} + \frac{\partial p}{\partial t} = 0$$

also with $dx = V\, dt$ and $V = 1/\sqrt{\varepsilon\varepsilon_0\mu\mu_0}$

From the second equation of (7.3.3) and from (7.3.2), we obtain:

$$\frac{1}{V^2}\frac{\partial E_r}{\partial t} = -\frac{\partial B_\theta}{\partial x}$$

If we still have $dx = V\, dt$, from this we obtain $E_r = -vB_\theta$.

By integrating along the vertical of the conductor

$$\int_h^{r_0} E_r\, dr = U_{\text{cond}} - U_{\text{ground}}$$

$$\int_h^{r_0} B_\theta\, dr = A_{x\,\text{cond}} - A_{x\,\text{ground}}$$

we then obtain

$$\left(U_{\text{ground}} \cong 0\right) \quad \text{and} \quad U_{\text{cond}} = -v\left(A_{x\,\text{cond}} - A_{x\,\text{ground}}\right)$$

As we saw that $J = QV$, and moreover since $U = \Delta Q$, finally we obtain:

TYPES OF MODELLING USED 147

$$A_{x\,cond} - A_{x\,ground} = \frac{1}{V^2}\Delta J \qquad (7.3.10)$$

When there is more than one conductor, the hypothesis of symmetry $\partial A_x/\partial \theta = 0$, perpendicular to each conductor, can only be admitted if the distances between conductors are small in relation to the mean heights of the latter from the ground, which is generally the case. One can therefore also write:

$$\overline{A}_{x\,cond} - \overline{A}_{x\,ground} = \frac{1}{V^2}|\Delta|\overline{J} \qquad (7.3.11)$$

in which $|\Delta|$ is the matrix of the electrostatic induction coefficients (cf. Figure 7.2).

We therefore have the following relationships between the potentials (U,A) at the surface of the conductors and of the ground, and the charges and currents in these conductors:

$$\begin{cases} \overline{U} = |\Delta|\overline{Q} \\ \overline{A}_{x\,cond} - \overline{A}_{x\,ground} = \frac{1}{V^2}|\Delta|\overline{J} \end{cases}$$

To pass to the circuit formulation where there is only $\overline{U}$ and $\overline{J}$, $\overline{Q}$ and $\overline{A}$ are eliminated by using the first equation of (7.3.6), Ohm's law (7.3.2) and the conservation of charges (7.3.4).

The following is derived from (7.3.6):

$$E_x = -\frac{\partial U}{\partial x} - \frac{\partial A_x}{\partial t} \qquad (7.3.12)$$

As the conductive materials are linear, for each harmonic state with frequency ω one can define a complex transmittance linking the longitudinal electric field E_x at the surface of the phase conductor (E_x is independent of θ) to the total current J flowing through it, that is $E_{x\,cond} = ZJ$, in which Z is the internal impedance of the conductor, varying with ω (induction effect and skin effect).

At the surface of the ground, the field E_x varies along a direction perpendicular to that of the line. One can however consider a reference $E_{x\,ground}$ at a point situated perpendicular to the phase conductor and write: $E_{x\,ground} = Z_0 J_0$, in which Z_0 is the apparent internal impedance of the ground and in which $J_0 = -J_{cond}$.

When there is more than one phase conductor, knowing that $U \cong 0$ at the ground surface, (7.3.12) is written as follows:

148 ELECTROMAGNETIC TRANSIENTS

$$E_{xm} = -\frac{\partial A_{xm}}{\partial t} = -\sum_n z_{omn} J_n$$

in which E_{xm} is the field at the ground surface perpendicular to the conductor m, J_n are the currents in the phase conductors and z_{omn} the apparent mutual impedances between the ground and the conductors. In fact, when the distances between phase conductors are small in relation to their height from the ground, the values of z_{omn} are close to the dominant value z_{dom} and we find again a mean value of the field E_{xm} : $E_{x\,ground} = Z_0 J_0$, in which $J_0 = -\Sigma_n J_n$ is the zero-phase sequence current.

Equation (7.3.12) is then written, using (7.3.11), for each phase conductor m:

$$\frac{\partial U_m}{\partial_x} = -Z_m J_m - \frac{p}{V^2}\sum_n \Delta_{mn} J_n - \sum_n z_{0mn} J_n$$

which is, in a matrix form:

$$\frac{\partial \overline{U}}{\partial x} = -\mathbf{Z}\overline{\mathbf{J}} \qquad (7.3.13)$$

with

$$Z_{mm} = z_m + z_{omn} + \frac{p}{V^2}\Delta_{mn} \quad \text{if } m = n$$

$$Z_{mn} = z_{omn} + \frac{p}{V^2}\Delta_{mn} \quad \text{if } m \neq n$$

The values of Δ_{ij} are independent of the frequency in the band (0–10 kHz) where our hypotheses are situated. The terms z vary with frequency, in a manner not related on account of the skin effect (perceptible at these frequencies particularly in the ground).

The conservation of charges (7.3.4) is written here for each conductor m:

$$\frac{\partial J_m}{\partial x} + \frac{\partial Q_m}{\partial t} + J_{m\,\text{leakage}} = 0$$

The current $J_{m\,\text{leakage}}$ for the conductor m is, as a first approximation, a function of the surface field of this conductor, and hence a function of its surface charge Q_m. This dependence is, in a first linear approximation: $J_{m\,\text{leakage}} = g_m Q_m$. In this context, one can take identical values of g_m if the phase conductors are identical and independent of the frequency. However, for underground cables, g_m can vary with frequency, since the losses in the dielectric are variable.

If we ignore the corona effect which introduces dependence of g_m on the voltage above a certain threshold and makes the problem nonlinear, one can then write:

$$\frac{\partial \bar{J}}{\partial x} = -Y\bar{U} \tag{7.3.14}$$

in which

$$Y_{ij} = (p+g_i)\Delta_{ij}^{-1}$$

Equations (7.3.13) and (7.3.14) are called the 'telegram operators' equations. To solve these, it is important to calculate the matrices **Z** and **Y** in the frequency band considered.

For overhead lines, g_i, which in any case is very low and difficult to measure, is often ignored (its value would be zero in the perfect dielectric model).

Calculation of the matrix **Y** is simple:

$$Y \cong p\Delta^{-1}$$

in which Δ_{ij} is given by the formulae of Figure 7.2.

To calculate Z, we already have the terms

$$\frac{P}{V^2}\Delta_{ij}$$

The z_m and z_{omn} still remain to be calculated.

Calculating the internal impedance of phase conductors

The flow-lines of current and the field E are parallel to the direction of the line inside the conductors and the ground. By taking the curl of the first equation (7.3.3), we have:

$$-\nabla^2 E = -p\left[\mu\mu_0 j + \frac{p}{V^2}E\right]$$

which, by using (7.3.2), becomes:

$$-\nabla^2 j - \frac{p^2}{V^2}j - p\sigma\mu\mu_0 j = 0 \quad \text{and} \quad \mu_0\varepsilon_0\varepsilon_2 = \frac{1}{V^2}$$

$$-\nabla^2 j - \frac{1}{V^2}\left[p^2 j + \frac{p\sigma}{\varepsilon\varepsilon_0}j\right] = 0 \tag{7.3.15}$$

j being parallel to the axis of the line, we obtain in cylindrical coordinates:

$$\frac{\partial^2 j}{\partial r^2} + \frac{1}{r}\frac{\partial j}{\partial r} + \frac{1}{r^2}\frac{\partial^2 j}{\partial \theta^2} + \frac{\partial^2 j}{\partial x^2} - \frac{p^2}{V^2}\left(1 + \frac{\sigma}{\varepsilon\varepsilon_{0p}}\right)j = 0$$

Now, it has been seen that the time constant $\varepsilon\varepsilon_0/\sigma$ was of the order of 10^{-19} s for the phase conductors, therefore $1 \ll \sigma/\varepsilon\varepsilon_{0p}$. For a line without losses

$$\frac{\partial^2 j}{\partial x^2} = \frac{p^2}{V^2} j$$

Now if

$$\frac{p^2}{V^2} j \ll \frac{p^2 \sigma j}{\varepsilon\varepsilon_{0p}}$$

one can also ignore $\partial^2 j/\partial x^2$ in front of $p^2\sigma j/\varepsilon\varepsilon_{0p}$. We therefore have:

$$\frac{\partial^2 j}{\partial r^2} + \frac{1}{r}\frac{\partial j}{\partial r} + \frac{1}{r^2}\frac{\partial^2 j}{\partial \theta^2} - \mu\mu_0 \sigma p j = 0 \qquad (7.3.16)$$

The solutions to (7.3.16) can take the following form:

$$J = \sum_{n=0}^{\infty} a_n I_n r(\mu\mu_0 \sigma p)^{1/2} \cos n\theta$$

in which I_n is the modified Bessel function of the first type of order n.

The space harmonics $h \geq 1$ having a negligible influence on the Joule losses under the approximation of low damping, only term a_0 is taken into account.

The total current J is obtained by integration over a cross-section:

$$J = \iint j \, dS = \int_0^{r_0} 2\pi r a_0 I_0 r(\mu\mu_0 \sigma p)^{1/2} \, dr$$

$$= \frac{2\pi a_0 r_0}{(\mu\mu_0 \sigma p)^{1/2}} I_1 r_0 (\mu\mu_0 \sigma p)^{1/2}$$

as

$$\int_{z_0}^{z_1} z^\gamma I_{\gamma-1}(z) \, dz = \left[z^\gamma I_\gamma(z) \right]_{z_0}^{z_1}$$

(property of the functions I_n).

Therefore the fundamental of the current in a conductor varies with the distance r to the axis of this conductor, according to:

$$j_0(r) = \frac{\sqrt{\mu\mu_0 \sigma p}}{2\pi r_0} \frac{I_0\left(r\sqrt{\mu\mu_0 \sigma p}\right)}{I_1\left(r\sqrt{\mu\mu_0 \sigma p}\right)} J.$$

TYPES OF MODELLING USED 151

Therefore, according to (7.3.2):

$$E_x(r_0) = \frac{1}{2\pi r_0}\left(\frac{\mu\mu_0 p}{\sigma}\right)^{1/2} \frac{I_0 r(\mu\mu_0 \sigma p)^{1/2}}{I_1 r(\mu\mu_0 \sigma p)^{1/2}} J$$

Hence

$$Z_m = \frac{1}{\pi r_0^2 \sigma}\frac{r_0\sqrt{\mu\mu_0\sigma p}}{2}\frac{I_0\left(r_0\sqrt{\mu\mu_0\sigma p}\right)}{I_1\left(r_0\sqrt{\mu\mu_0\sigma p}\right)} \qquad (7.3.17)$$

For $p = 0$, $I_0 = 1$ and $I_1 \cong \frac{1}{2}r_0(\mu\mu_0\sigma p)^{1/2}$ and we find again the resistance on direct current $1/\pi r_0^2\sigma$. At high frequencies, I_0 and I_1 have the same asymptotic development $I(z) \approx e^z/(2\pi z)^{1/2}$ and therefore:

$$z_m \cong \frac{1+j}{2\pi r_0}\frac{(\mu\mu_0\omega)^{1/2}}{2\sigma}$$

for $p = j\omega$ and $\omega \to \infty$.
By varying $j\omega$ and taking

$$R = \mathrm{Re}(z_m) \quad \text{and} \quad L = \frac{1}{\omega}\mathrm{Im}(z_m)$$

Figure 7.3 makes it possible to show the pattern of variations of R and L with frequency for the zero-phase sequence mode of a line (therefore making more use of the terms z_{0mn}).

Calculation of the apparent internal impedance of the ground

It has been seen above that the values of z_{0mn} are not very different from the same value z_0 for a polyphase line, the transverse dimensions of which are smaller than the mean height of the conductors in relation to the ground. Let us calculate z_0 in the case of a single-phase line above the ground.
In the ground, we still have:

$$\frac{\partial^2 j}{\partial x^2} + \frac{\partial^2 j}{\partial y^2} + \frac{\partial^2 j}{\partial z^2} - \frac{p^2}{V^2}\left(1+\frac{\sigma}{\varepsilon\varepsilon_{0p}}\right)j = 0$$

In a plane passing through the conductor and perpendicular to the ground $\partial^2 j/\partial y^2 \cong 0$.

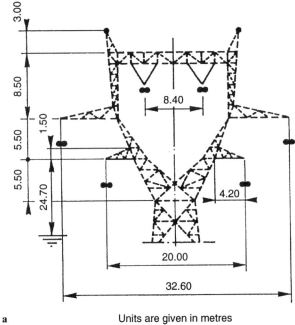

a

Units are given in metres

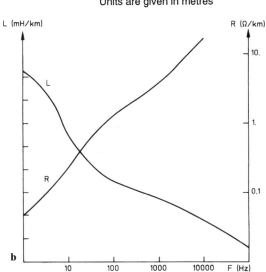

b

Fig. 7.3 Pattern of the variation R(ω) in log-log scale and L(ω) in semi-log scale, for the zero-phase sequence mode of a 400 kV line.

In addition, for the same reasons as above:

$$\frac{p^2}{V^2}j \ll \frac{\sigma p^2 j}{V^2 \varepsilon_0} \quad \text{and thus equally} \quad \frac{\partial^2 j}{\partial x^2} \ll \frac{\sigma p^2 j}{V^2 \varepsilon_0}$$

Therefore:

$$\frac{\partial^2 j}{\partial z^2} - \mu\mu_0 \sigma p j = 0 \quad \text{hence} \quad j(z) = j_0 e^{z\sqrt{\mu\mu_0 \sigma_p}} \tag{7.3.18}$$

Instead of concentrating on the surface, the current density is exponentially distributed in the ground e^{z/h_c} (z is negative) and $h_c = 1/(\mu\mu_0 \sigma p)^{1/2}$.

For a frequency of 10^4 Hz, $\sigma_{\text{ground}} = 10^{-2}$, we have $h_c = 35.6$ m and at 50 Hz $h_c \cong 503$ m.

Assuming a uniform current in the skin thickness h_c of the ground, one can approximately calculate the magnetic field $\boldsymbol{B}$ at the surface of the ground by:

$$j_{\text{ground}} = \frac{1}{\mu\mu_0}\left(\frac{\partial B_y}{\partial z} - \frac{\partial B_z}{\partial y}\right)$$

perpendicular with the conductor $\partial B_z/\partial y \cong 0$ and B_y has the same variation as a function of z as $j(z)$, equation (7.3.8), therefore $\partial B_y/\partial z = (\mu\mu_0 \sigma p)^{1/2} B_y$.

Now, perpendicular to the conductor

$$B_y = \frac{\mu\mu_0 J}{2\pi}\left(\frac{1}{h} + \frac{1}{h_c}\right)$$

Therefore

$$J_0 = \frac{1}{2\pi}\left(\frac{1}{h} + \frac{1}{h_c}\right)(\mu\mu_0 \sigma p)^{1/2}$$

at the surface of the ground. Now, $E_{x\,\text{ground}} = j_0/\sigma = z_0 J$. Therefore

$$z_0 = \frac{1}{2\pi}\left(\frac{1}{h} + \frac{1}{h_c}\right)\left(\frac{\mu\mu_0 p}{\sigma}\right)^{1/2}$$

It has been seen that generally $h_c \gg h$, hence

$$z_0 \cong \frac{(1+j)}{2\pi h}\left(\frac{\mu\mu_0 \omega}{2\sigma}\right)^{1/2} \tag{7.3.19}$$

for a harmonic state with angular frequency ω.

Theory of 'natural' modes

The two equations known as 'telegram operators' equations have been obtained as Laplace variables:

$$\begin{aligned} \frac{dU}{dx} &= -ZJ \\ \frac{dJ}{dx} &= -YU \end{aligned} \qquad (7.3.20)$$

with

$$Z_{mm} = z_m + z_{omn} + \frac{p}{V^2}\Delta_{mm}$$

$$Z_{mm} = z_{omn} + \frac{p}{V^2}\Delta_{mn}$$

and

$$Y_{ij} = (p + g_i)(\Delta^{-1})_{ij}$$

We have previously seen how to calculate, as a function of $p = j\omega$, the terms z_n, z_{omn} and Δ_{ij}, the latter not varying with ω.

Taking frequency as a variable, we are thus led to solve two ordinary matrix differential equations, linking the macroscopic variables U and I, thus making it possible to interface this model easily with models of system components having lumped elements, such as transformers and alternators.

By derivation from (7.3.20), we obtain:

$$\begin{cases} \dfrac{d^2U}{dx^2} = ZYU \\ \dfrac{d^2J}{dx^2} = YZJ \end{cases} \qquad (7.3.21)$$

For each frequency, one can define the normal of voltages and currents, transformations which diagonalize the matrices ZY and YZ with the respective eigenvalues γ_u and γ_i. In practice, the values γ_u and γ_i are very close together. In fact, if the values of g_i are all equal:

$$ZY = (p+g)\left[Z\Delta^{-1} + \frac{p}{V^2}\right]$$

$$YZ = (p+g)\left[\Delta^{-1}Z + \frac{p}{V^2}\right]$$

TYPES OF MODELLING USED 155

The matrix of the internal impedances **Z** and those of the influencing coefficients |Δ| have the same symmetries and therefore the same normal directions. One can therefore write:

$$\gamma^2 = K^{-1}ZYK = C^{-1}YZC$$

K being the transfer matrix to natural modes of voltage, **C** being the transfer matrix to natural modes of current and **γ** being the diagonal matrix of the model propagation constants.

One can then solve (7.3.21) by writing the conditions at the limits on U, I, dU/dx and dI/dx at one of the ends of the line.

The following relationships are then obtained between the voltages and currents at the ends 1 and 2:

$$\begin{pmatrix} U_2 \\ I_2 \end{pmatrix} = \begin{bmatrix} [\operatorname{ch}\gamma\ell][\operatorname{sh}\gamma\ell]\cdot[Z_c] & 0 \\ -[\operatorname{sh}\gamma\ell][Z_c]^{-1} & -[\operatorname{ch}\gamma\ell] \end{bmatrix} \begin{pmatrix} U_1 \\ I_1 \end{pmatrix} \qquad (7.3.22)$$

in which $Z_c = \gamma^{-1}KZC^{-1}$ is the matrix of the modal characteristic impedances, which can be shown to be diagonal.

Around the power frequency of 50 Hz or 60 Hz, it is then seen to be possible to replace the line defined by (7.3.22) by an equivalent π circuit formed by constant impedances, as shown on Figure 7.4.

An equivalent circuit can be deduced from this, consisting of lumped elements R, L and C which provide a line model in the form of differential equations in relation to time, valid in a small frequency band about the industrial frequency. This model is sufficient, for example, to take into account overvoltages under steady-state conditions, such as those gener-

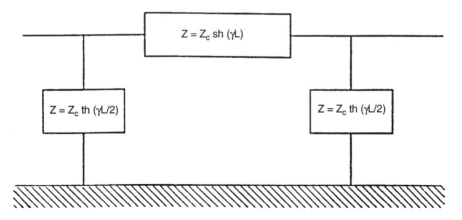

Fig. 7.4 Equivalent π circuit of the line near the power frequency of 50 Hz.

ated by the Ferranti effect. This is the phenomenon of natural multiplication of the voltage at the open end of a long line, in relation to the voltage of the source connected at the other end. Let us turn to equation (7.3.22) again and take $I_2 = 0$ (open end). We have:

$$U_2 = (\operatorname{ch} \gamma \ell) U_1 + (\mathbf{Z}_c)(\operatorname{sh} \gamma \ell) I_1$$
$$0 = (-\operatorname{sh} \gamma \ell)(\mathbf{Z}_c^{-1}) U_1 - (\operatorname{ch} \gamma \ell) I_1$$

for each of its modes. Hence $U_2 = U_1 / \operatorname{ch}(\gamma \ell)$.

If the damping is fairly low, we have

$$\frac{1}{\operatorname{ch} \gamma \ell} \approx 1 - \frac{\gamma^2 \ell^2}{2}$$

In comparison with the propagation without loss in which $\gamma = j\omega(LC)^{1/2}$ we have here $\gamma = \alpha + j\beta$ in which $\beta \gg \alpha$ in the case of low losses. Therefore

$$\frac{U_2}{U_1} \cong 1 + \frac{\beta^2 \ell^2}{2}$$

ℓ being the length of the line. This formula makes it possible to find, on a line on no-load 300 km long, the permanent overvoltage of 5% at the open end.

Return to the time domain for the simulation of transients

To obtain the dynamics of the lines in a wider frequency band, it is necessary to consider the wave variables. Let us consider the case of a line without losses in which the resolution by separation of variables reveals wave quantities $U \pm Z_c J$; the two equations of (7.3.22) can be combined to obtain:

$$\begin{aligned}(U_2 + Z_c I_2) &= e^{-\gamma \ell}(U_1 - Z_c I_1) \\ (U_1 + Z_c I_1) &= e^{-\gamma \ell}(U_2 - Z_c I_2)\end{aligned} \qquad (7.3.23)$$

Z_c and γ are calculated at each frequency from the matrices $|\mathbf{Z}|$ and $|\mathbf{Y}|$ of (7.3.20), from which the inherent values of the product are sought. Having completed this calculation, one can go back to the time domain by carrying out an inverse Fourier transform.

The transfer matrices are deemed to be constant, which is verified for the lines in the band [0–10 kHz].

The simple product is then expressed by the convolution denoted by *
and we obtain:

$$u_2(t) + z_c * i_2(t) = r(t) * (u_1(t) - z_c * i_1(t))$$
$$u_1(t) + z_c * i_1(t) = r(t) * (u_2(t) - z_c * i_2(t))$$
(7.3.24)

in which $z_c(t) = F^{-1}(Z_c(\omega))$ and $r(t) = F^{-1}(e^{-\lambda(\omega)\cdot \ell})$.

The treatment is different for the overhead modes and for those which involve preparation in the ground (zero-phase sequence and intercircuit modes for soluble circuit lines).

For the overhead modes, we find that Z_c does not vary much with frequency, except when nearby direct current. The pulse response of the line $r(t)$ appears as a pure time delay with constant damping for these modes.

If the index α designates overhead modes, on the basis of (7.3.24) one can write:

$$u_2^\alpha(t) + z_0^\alpha i_2^\alpha(t) = \lambda^\alpha \left[u_1^\alpha(t - \tau^\alpha) - z_0 i_1^\alpha(t - \tau^\alpha) \right]$$
$$u_1^\alpha(t) + z_0^\alpha i_1^\alpha(t) = \lambda^\alpha \left[u_2^\alpha(t - \tau^\alpha) - z_0 i_2^\alpha(t - \tau^\alpha) \right]$$

in which z_0^α is the surge impedance of the overhead modes,
 λ^α is the damping ($= 1 - \varepsilon$) of these modes and
 τ^α is their propagation time delay $\cong (\ell/c)$.

For the zero-phase sequence mode, $r(t)$ still causes a time delay to appear; however, a much greater dispersion about the group delay is observed than for the overhead modes.

Figure 7.5 illustrates this phenomenon. Modal weighting functions of the overhead and of the zero-phase sequence mode for a three-phase line are shown.

For the zero sequence mode, it is also important to take into account the variation Z_c with frequency. For this mode it is therefore necessary to evaluate all the products of convolutions of the system (7.3.24).

Whatever the case may be, we note that on account of the time delay present in $r(t)$, the resolutions of the wave variables at ends 1 and 2 of a line are decoupled. This comment forms the basis of the method of calculation used in electromagnetic transient programs, permitting integration of the line model into lumped-element models without additional expense. This even makes it possible to accelerate the solution of the overall system, since two subsystems linked only by lines can be dealt with independently and in parallel.

This model makes it possible to calculate the transient response of lines

158 ELECTROMAGNETIC TRANSIENTS

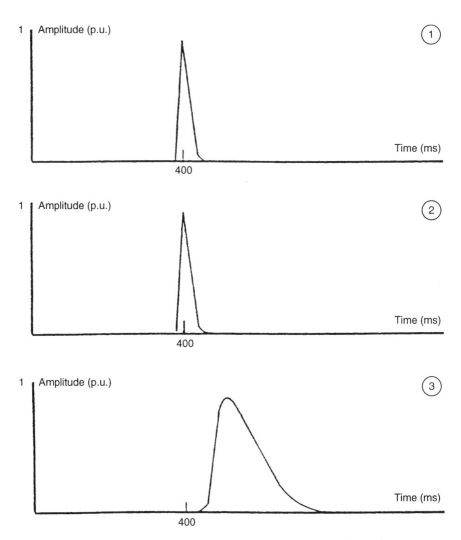

Fig. 7.5 Modal pulse response of a three-phase line. Curves 1 and 2 = overhead modes; curse 3 = zero-phase sequence mode (p.u. = per unit).

in the frequency band initially specified (0–10 kHz) and therefore allows us to take into account the phenomena mentioned in section 7.1.

As an illustration of this, Figure 7.6 shows the transient response of a line at the moment when it is energized. An overvoltage of almost 100% can be generated at the open end during this type of switching operation.

The polarity reversal on phases 2 and 3 at the first wavefront shows that

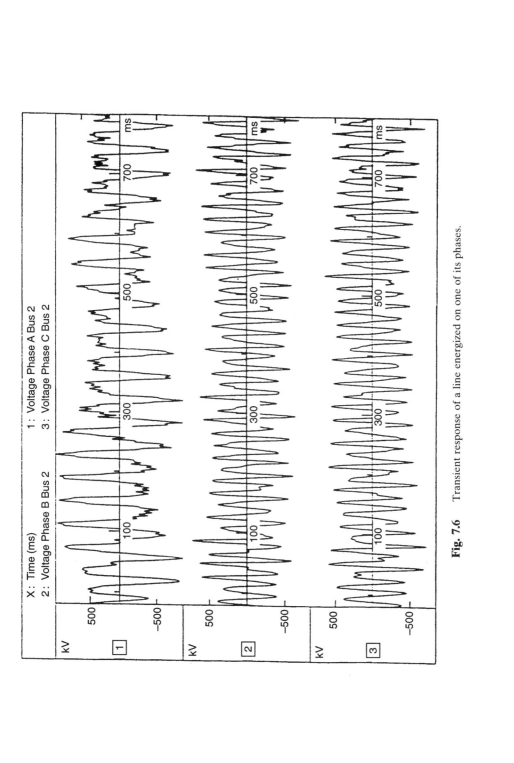

Fig. 7.6 Transient response of a line energized on one of its phases.

this is an overhead mode (with zero sum). The second front, of identical polarity, is the zero sequence mode, which therefore appears to be slower than the overhead modes.

7.3.2 Modelling transformers

Transformers are one of the examples of devices modelled by lumped-element discrete components. For the frequencies of interest to us, it is not necessary to represent propagation in transformers. In contrast, to account for the internal resonances which can appear in transformers, it will sometimes be necessary to make the model of the windings slightly more complex, by including capacitance between turns or between the winding and the tank, etc.

A simplified model is presented here, valid up to a few kilohertz, but taking into account one of the essential phenomena, which is the saturation of the magnetic circuits. This model has proven to be sufficient to represent the ferroresonance phenomena mentioned in section 7.2.

The transformer is considered as a group of windings enclosing magnetic circuits of various shapes (Figure 7.7).

The three-phase windings can be connected in several possible ways: star, delta, zigzag, ... (Figure 7.8).

These methods of connection of three-phase windings basically serve to obtain the given characteristics under zero-phase sequence conditions. We take as a basis the Maxwell expression (7.3.3), in which the displacement

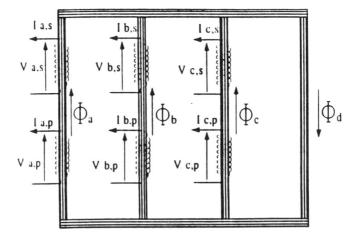

Fig. 7.7 Diagram of a transformer with two three-phase windings, with free flux.

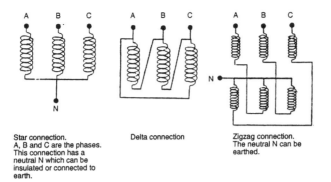

Fig. 7.8 Examples of connections of three-phase transformer windings.

current $\partial D/\partial t$ is ignored. Their integral expression, using Stokes's formula, gives Lenz's law and Ampere's theorem.

$$e = \frac{d\varphi}{dt}$$
$$\oint H\,dl = \Psi \qquad (7.3.25)$$

in which φ = magnetic induction vector flux,
Ψ = current density vector flux.

On account of the values of relative permeability μ_r of the magnetic materials, certain magnetic field lines are not channelled and close up in the vacuum. Iron channels induction lines less well than copper channels current lines, hence the greater leakage than in electric conductors.

Assuming magnetic circuits of low curvature and constant cross-section, one can define reluctance of the elements of the magnetic circuit. By applying the second equation of (7.3.25):

$$\oint H\,dl \cong \boldsymbol{H} \cdot \ell = ni \qquad (7.3.26)$$

where ℓ is the mean length of the line of the magnetic circuit and $\boldsymbol{H}$ is constant in the magnetic material providing that the cross-section remains constant.
From

$$\boldsymbol{B} = \mu_0 \mu_r \boldsymbol{H}$$
$$\Phi = \boldsymbol{B} \cdot \boldsymbol{S}$$

in which $\boldsymbol{S}$ is the cross-section of the magnetic circuit and from (7.3.26) we obtain

$$ni = \Re\Phi \qquad (7.3.27)$$

in which n is the number of turns through which the lines of field H passes, and in which $\Re = \ell/\mu_0\mu_r S$, which is only dependent on the circuit geometry, is the reluctance of the magnetic circuit.

Formula (7.3.27) makes it possible to deal with the magnetic circuit using a formal method analogous to the Kirchhoff laws for electrical circuits. In fact ni, called the magnetomotive force, is analogous to electromotive force, $\Re$, reluctance, is analogous to electrical resistance and Φ is analogous to the current. The law (7.3.27), the analogue of Ohm's law, allows Kirchhoff's second law to be written for any magnetic circuit. Kirchhoff's first law arises from the fact that the sum of the fluxes (with their sign) at a mode is zero, a direct consequence of $\nabla \cdot \boldsymbol{B} = 0$.

Since the flux is imperfectly channelled by the magnetic material, it is interesting to consider the magnetic leakages separately (Figure 7.9).

The flux carried by the iron is associated with a reluctance $\Re_{iron} = \ell/\mu_0\mu_r S$, and the leakage flux is associated with a virtual reluctance $\Re_{leakage} \gg \Re_{iron}$ in parallel with the preceding one. One can therefore write:

$$ni = \Re_{iron}\Phi_{iron} = \Re_{leakage}\Phi_{leakage}$$

By using equations (7.3.25), a circuit like that in Figure 7.9 can then be put into equations:

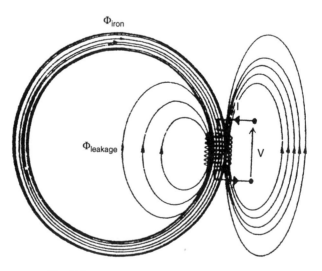

Fig. 7.9 Representation of magnetic leakages.

$$\vartheta = r_i + n\frac{d\Phi}{dt}$$

$$\Phi = \Phi_{\text{iron}} + \Phi_{\text{leakage}}$$

$$ni = \Re_{\text{iron}}\Phi_{\text{iron}} = \Re_{\text{leakage}}\Phi_{\text{leakage}}$$

hence:

$$\vartheta = r_i + \frac{n^2}{\Re_{\text{leakage}}}\frac{di}{dt} + \frac{n^2}{\Re_{\text{iron}}}\frac{di}{dt} \qquad (7.3.28)$$

$n^2/\Re_{\text{leakage}}$ is the leakage inductance and $n^2/\Re_{\text{iron}}$ is the normal inductance.

In the event of the magnetic material being saturated, the main hypothesis made consists of assuming that $n^2/\Re_{\text{leakage}}$ is constant and that only the flux in the iron is saturating.

This hypothesis is valid only for slightly saturated states. In states with heavy saturation, the pattern of the field **H** changes and the leakage inductance tends to increase as iron carries the magnetic flux less well. In everything which follows, we shall however continue to assume that $n^2/\Re_{\text{leakage}}$ is constant, since conditions with low saturation correspond well to the operating states of the transformers, even exceptional ones, and since the modelling which arises from this hypothesis is more than adequate to simulate phenomena such as the ferroresonance or the inrush currents discussed in section 7.2.

Let us begin with the saturation curve of the magnetic material giving induction **B** as a function of magnetic field **H** (Figure 7.10).

From this, by simply changing the scale, one can deduce a curve (Φ, I) of

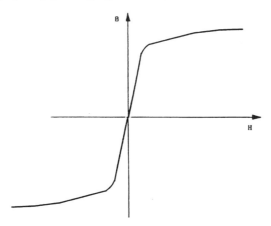

Fig. 7.10 Saturation curve of magnetic materials (first magnetization curve).

each homogeneous subassembly of the magnetic circuit (partial magnetization characteristic).

In fact $\Phi = B \cdot S_{useful}$, in which S_{useful} is the 'useful' cross-section (the part of the cross-section which effectively carries the flux), and $I = Hl/n$.

In the iron, $\nabla \wedge H = 0$, therefore the field H derives from a magnetic potential γ. For each part AB of the magnetic circuit, one can therefore write:

$$\gamma_A - \gamma_B = \Re(\Phi_{iron}) \cdot \Phi_{iron} - ni \qquad (7.3.29)$$

From the curve ($\Phi \cdot I$) one can easily deduce ($\Re, \Phi$) since $\Re = nI/\Phi$. Formula (7.3.29) therefore gives the partial magnetization characteristic of any homogeneous part of the magnetic circuit.

One can therefore express as an equation, keeping in mind the hypothesis simplifying the constant stray inductances, the three-phase transformer with two windings and free flux of Figure 7.7:

$$\begin{bmatrix} U^p \\ U^s \end{bmatrix} = \begin{bmatrix} r^p & 0 \\ 0 & r^s \end{bmatrix} \begin{bmatrix} I^p \\ I^s \end{bmatrix} + \begin{bmatrix} \ell^p & m \\ m & \ell^s \end{bmatrix} \frac{d}{dt} \begin{bmatrix} I^p \\ I^s \end{bmatrix} + \begin{bmatrix} n^p & 0 \\ 0 & n^s \end{bmatrix} \frac{d}{dt} \begin{bmatrix} \Phi \\ \Phi \end{bmatrix} \qquad (7.3.30)$$

in which U^p, U^s, I^p, I^s and Φ are three-phase vectors, n^p, n^s, r^p and r^s are diagonal matrices of the third order, ℓ^p, m and ℓ^s are 3 × 3 matrices, characterizing constant leakage inductances, and m can represent leakage mutual coefficients between primary and secondary windings.

For the magnetic circuit, by applying (7.3.29) and stating $\Phi_D = \Phi_a + \Phi_b + \Phi_c$ we have:

$$\begin{bmatrix} \Re_a(\Phi_a) + \Re'(\Phi_D)\Re'(\Phi_D)\Re'(\Phi_D) \\ \Re'(\Phi_D)\Re_b(\Phi_b) + \Re'(\Phi_D)\Re'(\Phi_D) \\ \Re'(\Phi_D)\Re'(\Phi_D)\Re_c(\Phi_c) + \Re'(\Phi_D) \end{bmatrix} \Phi = n^p I^p + n^s I^s \qquad (7.3.31)$$

The system of equations (7.3.31) is nonlinear. However, the matrix of the left member (whatever the vector Φ) is a M-matrix which can be formally inverted, which makes it possible to eliminate Φ between equations (7.3.30) and (7.3.31) and obtain a system of nonlinear differential equations linking the primary and secondary three-phase voltages U^p and U^s with the three-phase currents flowing in the windings, I^p and I^s, which is perfectly consistent with the formal method used to model the other elements of the system such as lines or alternators; see section 7.3.3.

Once these differential equations have been discretized in time, by using an implicit integration scheme, they imply the solution of a nonlinear system at each time step, which is generally achieved by using one of Newton's methods.

The eddy currents are taken into account by simulating the losses which they cause, which are proportional to the square of the voltage amplitude. These are therefore represented by a constant shunt resistance on the primary and secondary terminals of the transformer. The hysteresis loss can also be modelled by a nonlinear resistance as a function of the root-mean-square value of the voltage, if we consider that the energy dissipated by hysteresis is proportional to the area of the main hysteresis loop. If we wish to represent secondary loops, a slightly more accurate model of behaviour is required. The model considered makes it possible to rejoin the upper or lower bounds of the main cycle, as a function of the sign of dB/dt, with a time constant, making possible secondary cycles simulation (Figure 7.11). This additional modelling does not change the formulation of (7.3.31) except that the reluctances no longer depend only on the instantaneous values of Φ, but also past values of Φ at the last instant of simultaneous passing through zero of the derivatives $d\Phi/dt$ and di/dt.

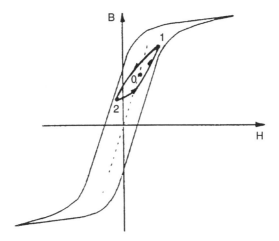

Fig. 7.11 Model of behavior of the hysteresis permitting simulation of secondary cycles: at the instant t, the reluctance is dependent on the instantaneous flux and the flux prior to the last cusp (2).

With this model the dynamic behaviour of transformers can be represented in a frequency band of several kilohertz, and even beyond this, if the modelling of the windings is made more complex by the addition of capacitances between turns and between windings and tank.

This model has proven to be sufficient to reproduce the behaviour in an unbalanced state, in a transient state (inrush currents) and in a nonlinear state (bifurcation towards ferroresonant states). As an illustration, Figure 7.12 shows a ferroresonant state appearing on inadvertent opening of one phase of a transformer.

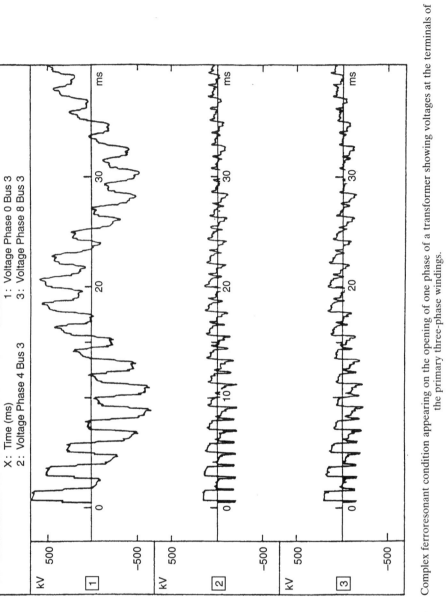

Fig. 7.12 Complex ferroresonant condition appearing on the opening of one phase of a transformer showing voltages at the terminals of the primary three-phase windings.

TYPES OF MODELLING USED   167

A second subharmonic state appears on the phase in question and the other phases show irregular modulation of their fundamental condition at 50 Hz or 60 Hz.

7.3.3 Representation of synchronous machines

The classic equations of Park are used, the basis of which has been explained previously (Chapter 6).

We should simply remember that the basic hypothesis is the sinusoidal distribution of the magnetomotive force (therefore the conductors) of the armature (stator). Then (in the absence of saturation), inductances are forms in simple terms: constants and cosines of the rotor winding angle in relation to the axis of one of the phases of the stator.

Park's transform then shows to advantage as it considerably simplifies the relationships between flux and current (in fact it diagonalizes the submatrix relative to the stator phases) and we obtain:

$$\begin{aligned}
w_0 \varphi_{kd} &= X_D I_{kd} + X_{ad}\left(I_{fd} + I_{kd} + I_d\right) \\
w_0 \varphi_d &= X_s I_d + X_{ad}\left(I_{fd} + I_{kd} + I_d\right) \\
w_0 \varphi_{fd} &= X_{ffd} I_{fd} + X_{ad}\left(I_{fd} + I_{kd} + I_d\right) \\
w_0 \varphi_q &= X_s I_q + X_{aq}\left(I_{fq} + I_{kq} + I_q\right) \\
w_0 \varphi_{fq} &= X_{ffq} I_{fq} + X_{aq}\left(I_{fq} + I_{kq} + I_q\right) \\
w_0 \varphi_{kq} &= X_Q I_{kq} + X_{aq}\left(I_{fq} + I_{kq} + I_q\right)
\end{aligned}$$

(7.3.32)

Generally speaking, the consequences of the establishment, during transient conditions, of eddy currents in the solid rotors of turbo-alternators are represented by adding a circuit in parallel to the classic damping circuit of axis q (variables ϕ_{fq}, I_{fq} ... of the preceding equations).

We should remember that in Park's system, Lenz's law for the stator is written as follows:

$$\begin{aligned}
V_d &= \frac{d\varphi_d}{dt} - \varphi_q \frac{d\theta}{dt} - R_a i_d \\
V_q &= \frac{d\varphi_q}{dt} + \varphi_d \frac{d\theta}{dt} - R_a i_q \\
V_0 &= -\frac{d\varphi_0}{dt} - R_a i_0
\end{aligned}$$

(7.3.33)

and for the other circuits:

$$V_{fd} = \frac{d\phi_{fd}}{dt} + r_{fd}i_{fd} \quad \text{inductor}$$

$$0 = \frac{d\phi_{fd}}{dt} + r_{fq}i_{fq} \quad \text{solid rotor}$$

$$0 = -\frac{d\phi_{kd}}{dt} + r_{kd}i_{kd} \quad \text{dampers}$$

$$0 = \frac{d\phi_{kq}}{dt} + r_{kq}i_{kq} \quad \text{dampers}$$

Moreover, the electric torque is expressed by:

$$C_e = \frac{3}{2}\left(\varphi_d i_q - \varphi_q i_d\right)$$

To take into account the saturation of the magnetic circuits, the equations (7.3.32) giving flux values as a function of currents are modified.

Several methods based on physical considerations are used; they all amount to adding a nonlinear term to the expressions (7.3.32). These become:

$$w_0\varphi_{kd} = X_D I_{kd} + w_0\varphi_{rd}$$
$$w_0\varphi_d = X_s I_d + w_0\varphi_{rd}$$
$$w_0\varphi_{fd} = X_{ffd} I_{fd} + w_0\varphi_{rd}$$
$$w_0\varphi_q = X_s I_q + w_0\varphi_{rq}$$
$$w_0\varphi_{fq} = X_{ffq} I_{fq} + w_0\varphi_{rq}$$
$$w_0\varphi_{kq} = X_Q I_{kq} + w_0\varphi_{rq}$$

in which $\varphi_r = \begin{vmatrix} \varphi_{rd} \\ \varphi_{rq} \end{vmatrix}$ and

$$w_0\varphi_{rd} = X_{ad}\left(I_{fd} + I_{kd} + I_d - I_{sd}\right)$$
$$w_0\varphi_{rq} = X_{aq}\left(I_{fq} + I_{kq} + I_q - I_{sq}\right)$$

This formulation assumes that the flux across a winding is equal to the sum of a joint air gap flux φ_r (which alone is saturating) and a leakage flux inherent in the winding. We therefore admit that there are no leakages common to two windings and that the leakage inductance remains constant in the saturated state.

As stated previously, there are several ways of calculating the saturation current

$$I_s \begin{vmatrix} I_{sd} \\ I_{sq} \end{vmatrix}$$

These are all based on the use of no-load characteristics of the flux (or e.m.f.) as a function of the current.

One possible method is as follows, assuming that I_s and the common flux ϕ_r are in phase.

When there is a shift of ϕ_r in relation to the axis d equal to α (Figure 7.13), we will have:

$$\|I_s\| = f_d(\|\varphi_r\|) \cos^2\alpha + f_q(\|\varphi_r\|) \sin^2\alpha$$

where the functions f_d and f_q are the characteristics of the direct axis and quadrature axis saturation, respectively.

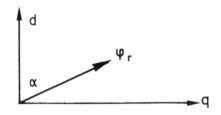

Fig. 7.13 Components of φ_r on p and q axes.

f_d is directly deduced from the open-circuit characteristic (giving the open-circuit e.m.f. as a function of the excitation current) as shown in Figure 7.14.

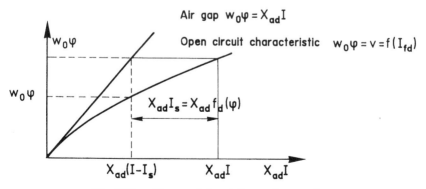

Fig. 7.14 Characteristics of direct axis saturation.

170 ELECTROMAGNETIC TRANSIENTS

Another experimental characteristic is used for the quadrature axis (and f_q).

In contrast to what happens when examining transient stability, here no simplification is introduced into the stator equations (7.3.33). We take into account the instantaneous speed of the rotor $d\theta/dt$ and the transformer electromotive forces $d\varphi/dt$.

At each step in time, the system of differential equations is integrated

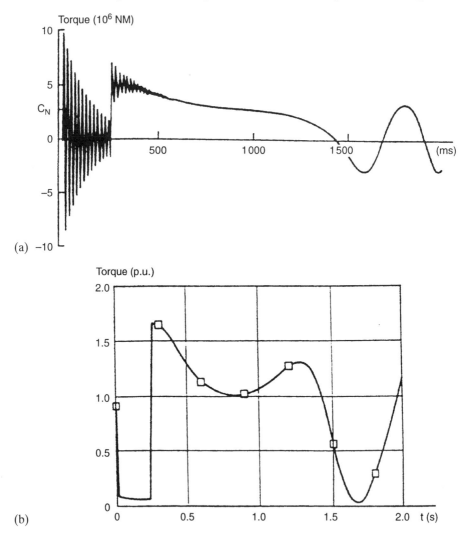

Fig. 7.15 Three-phase fault: electrodynamic torque fault with duration 245 ms and unit start-up time 10 s. (a) Electromagnetic transient simulation. (b) Electromechanical transient simulation.

and Park's inverse transform allows us to revert to the phase variables. The current and voltage waves following disturbances and the electrodynamics torque are thus reproduced accurately.

In particular, the 'back-swing' phenomenon is simulated: just after a short-circuit the electrodynamic torque increases and reaches several times the rated torque, and the machine slows down. Knowledge of this phenomenon is important when designing the machine and also in certain investigations into transient stability analysis, since torque has a direct influence on the angle and hence on stability. As an example, we have indicated below the variations in the electrodynamic torque and the angle of a machine obtained partly with the complete equations (electromagnetic transients program which we have just described), and with the classic simplified equations of transient stability; see Chapter 6.

Figures 7.15 and 7.16 show that the transient stability equations allow one to obtain a mean value of the torque which completely obscures the back-swing (which takes places over one half cycle). For low inertia machines this would lead to an error on the angle and stability of the machine, as seen in Figure 7.17.

We should also remember that taking into account the complete equations enables us to simulate short-circuit currents accurately with their DC

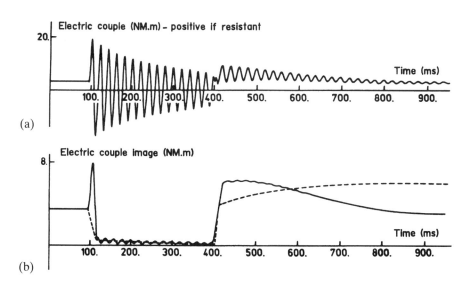

Fig. 7.16 Three-phase short-circuit cleared at $t = 400\,\text{ms}$: comparison of the electric torques obtained by simulation of electromagnetic transients and electromechanical transients. (a) Electromagnetic transient simulation. (b) Curve —: rolling average on 20 ms of the electrical torque calculated by using electromagnetic transient simulation; curve ----: electrical torque calculated by using electromechanical transient simulation.

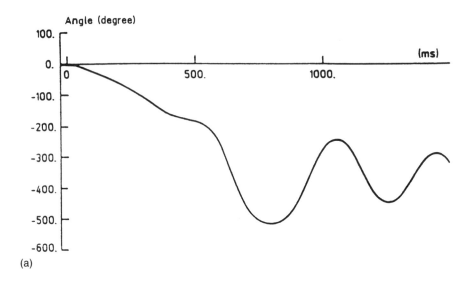

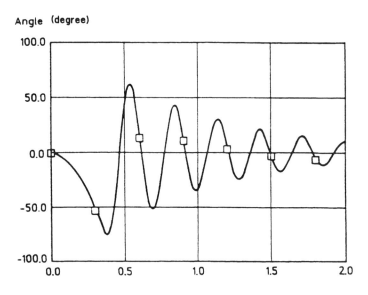

Fig. 7.17 Comparison of the angles obtained by simulation of electromagnetic and electromechanical transients. (a) Electromagnetic transient simulation with loss of synchronization. (b) Electromechanical transient simulation with resynchronization.

component, whilst equations of transient stability only allow us to obtain the AC component of the transient conditions (Figure 7.18).

There is no representation of voltage and speed regulations and of the excitation system dedicated to very high speed phenomena. The same representations are used as in transient stability.

The shaft line is modelled by a group of N masses linked by torsion springs (Figure 7.19).

Each spring is characterized by:

- stiffness $R_{i,i+1}$,
- a coefficient of viscous damping $K_{1,i+1}$.

Each mass is characterized by:

- a moment of inertia J_i,
- a coefficient of viscous friction F_i.

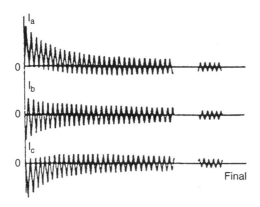

Fig. 7.18 Current variations in the three phases of the armature of a synchronous machine during a sudden three-phase fault.

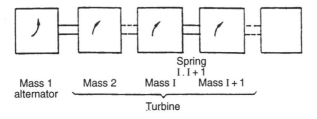

Fig. 7.19 Shaft line modelling of a turboalternator set.

If the angular position of the mass i is called θ_i, its velocity w_i, and the torque applied C_i, the equations are of the following type:

$$J_i \frac{dw_i}{dt} + R_{ri-1}(\theta_i - \theta_{i-1}) + R_{ri+1}(\theta_i - \theta_{i-1}) + F_i w_i$$
$$+ K_{ri-1}(w_i - w_{i-1}) + K_{ri+1}(w_i - w_{i+1}) = C_I$$

We can thus simulate the oscillations of the shaft line, and in particular the subsynchronous oscillations. These are caused by resonances between mechanical natural frequencies of the shaft line and electrical frequencies of the power system. To reproduce these therefore, in addition to the shaft line oscillations, the voltage and current waves carrying different frequency components must be simulated.

7.4 METHOD OF SOLUTION

We saw above how different components of the power system could be modelled for the application sought.

All these components, even nonlinear devices, with the exception of alternators, can be represented by equations in a form similar to those of Bergeron by integrating time derivatives with the trapezoidal rule.

By writing Kirchhoff's law on currents $\Sigma i(t) = 0$ at each node of the network, at each stage of the calculation we obtain a large real linear system in the form:

$$YV = I \qquad (7.4.1)$$

in which V is the vector of the instantaneous voltages at the different nodes, Y is an admittance matrix constructed from Bergeron's equations written for all the components of the power system, and I is the column vector of the currents 'injected', as a function of the previous states of the system only, thanks to the delay introduced by $r(t)$.

As we saw in section 7.3.1, the two ends of a line no longer 'see' one another directly, but through the intermediary of previously calculated terms. This decoupling of the equations plays an essential role in limiting the volume of calculations during simulation, by making it possible, at a given interval of time, to obtain an independent solution for each 'subsystem' linked to its neighbours only by lines.

The real matrix Y is very hollow, and dependent only on the topology of the power system. It should therefore be recalculated only each time the switchgear is operated during the simulated sequence, or in the presence of nonlinear elements (we should then use a Newton type method of resolution). Its size is three times the number of connection points between elements of the system.

We have thus brought a problem of partial derivative time and space equations down to a very classical method of solving large hollow real systems, generally linear, to determine V at each step of the calculation.

For alternators, we use Park's classic equations, as shown in section 7.3.3. At each interval of time, we integrate the system of differential equations of each machine separately (possibly representing the upstream turbine) and we apply Park's inverse transform to come back to the phase variables and to interface with the linear system (7.4.1).

In conclusion, Bergeron's method, associated with the integration of differential equations by the trapezoidal rule, is powerful and has permitted spectacular progress in modelling and the understanding of fast transient phenomena. The simulation of distributed elements may however still benefit from the progress in taking better account of distortion. This is necessary for the zero-phase sequence mode of overhead lines, and all the modes of propagation of underground cables, the use of which is increasing. In its simplified form (not taking distortion into account), Bergeron's method lends itself to very efficient computerized use. With a typical time step of $100\,\mu s$, and for an electrical system limited to a few lines and generating sets, around 150 000 instructions will be executed per interval, that is, approximately 1500 million instructions in one second of simulation. To represent nonlinearities and distortion (with the frequency dependence of parameters R and L), there will be around 1.5 million instructions per step of the calculations.

The success of Bergeron's method used by the majority of fast transient calculation tools should not mask certain reservations concerning its use. The approximations discussed in this chapter limit the validity of the simulations to a few tens of kilohertz, and it would be fruitless to carry out calculations at 500 kHz, let alone 1 Mz. Similarly, the geological structure of the ground is rarely that of a homogeneous and isotropic semi-space, which makes the quest for excessive precision in the simulation of propagation phenomena an illusion.

For very high frequencies, the only possible approach is still the three-dimensional solution of Maxwell's equations. A highly detailed representation of the components is then required (with numerous data) and calculation times become prohibitive as soon as we exceed a few system elements. Fortunately, we shall consider that, contrary to the case of research into the dimensioning of equipment, analysis of the electrical system in these frequency ranges is not truly necessary. We shall therefore not discuss it in this work.

FURTHER READING

Bergeron L. (1961). *Du Coup de Belier en Hydraulique au Coup de Foudre en Electricité*, Dunod, Paris (translated version: Wiley, New York).

Bornard P., Erhard P. and Fauquembergue P. (1988). 'Morgat': a data processing program for testing transmission time protective relays. *IEEE/PES*, **3**(4), October.

Budner A. (1970). Introduction of frequency-dependent line parameters into an electromagnetic transients program. *IEEE*, **PAS-89**, 88–97.

Dommel H.W. (1969). Digital computer solution of electromagnetic transients in single and multiple networks. *IEEE*, **PAS-88**, 388–99.

Meyer W.S. and Dommel H.W. (1974). Numerical modelling of frequency-dependent transmission line parameters in an electromagnetic transients program, *IEEE Winter Power Meeting*, New York, paper T74080-8.

Snelson J.K. (1972). Propagation of travelling waves on transmission lines: frequency dependent parameters. *IEEE*, **PAS-91**, 85–91.

8

HARMONICS

8.1 INTRODUCTION

At customers' mains, electrical energy systems are assumed to be three-phase sinusoidal voltage sources. However, in real life, this objective is never fully reached. During transmission, electricity inevitably deteriorates depending on its applications, thus causing a deformation of the sinusoids. This deformation or distortion is referred to as a harmonic disturbance.

Harmonic disturbance is due to a great extent to the development of loads fed by devices with a nonlinear current–voltage characteristic. These loads produce harmonic currents (or voltages) which propagate throughout the system, with the risk of enhancing the background noise of the system and engendering harmonic disturbances at all points. Their effects are only really felt from a certain threshold of amplitude, a threshold which can vary depending on the sensitivity of the equipment likely to be disturbed.

For an effective method to fight this disturbance, it is therefore indispensable:

- to identify and model harmonic sources;
- to predetermine, through computations, the levels of harmonic voltages generated depending on the topology of systems and the development of disturbing loads.

Nevertheless, harmonic studies are often confronted with difficulties linked to solving high-order equations representing the system as a whole formed of a large number of facilities and loads. Notwithstanding, the development of simulation methods allows large systems (400 nodes) to be modelled today and harmonic propagation conditions to be examined.

8.2 MODELLING OF SYSTEMS IN HARMONIC CONDITIONS

The modelling of systems in harmonic conditions involves two aspects:

- determining the impedance of the system to harmonic frequencies;
- representing the harmonic sources.

8.2.1 Determining harmonic impedances

The impedances of the elements of a system to harmonic frequencies are determined on the basis of their value at power frequency (50 Hz or 60 Hz).

The choice of a model depends on the accuracy sought, the availability of data and the range of frequencies considered. The majority are valid only for frequencies below a few kilohertz. Models of elements of the system are shown in Appendix 8.A.

The complete system can never be described in full in all harmonic studies (study of propagation, connection of disturbing equipment, study of filtering of an installation, etc.). It is therefore necessary to limit the dimensions of the system, for some parts, using the equivalent impedances representing its behaviour to harmonic disturbance.

These impedances vary to a great extent over time and from one point of the system to the next. They depend, among other things, on the short-circuit power of systems, the length of lines (operating schedules), the presence of banks of capacitors for reactive energy compensation (closed or opened during the day), and the load level of the system.

Either an experimental or a digital simulation method or a combination of both can be used to estimate these impedances.

8.2.1.1 Measurement of the harmonic impedance of systems

This type of measurement is quite difficult to implement. It necessitates the presence of a sufficiently powerful source of harmonic currents or a relatively high pre-existing harmonic voltage [1–3] at the node where impedance is to be measured.

Use of pre-existing (inter) harmonic voltages

The pre-existing (inter) harmonic voltage V_{h_1} in the system causes an (inter) harmonic current to flow in load Z (linear or nonlinear) at which boundaries a voltage V_{h_2} appears (Figure 8.1). The harmonic impedance Z_h is then written:

$$Z_h = \frac{V_{h_1} - V_{h_2}}{i_h}$$

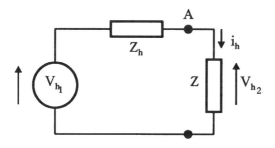

Fig. 8.1 Thevenin harmonic equivalent diagram of the system seen from measuring point A.

Use of an (inter) harmonic current source

Assuming that there is no (inter) harmonic voltage in the system before injection of harmonic currents produced by a disturbing unit of equipment (rectifier bridges, etc.) or through the operation of electrical equipment (transformer, bank of capacitors, etc.), this injection produces a harmonic voltage V_h (Figure 8.2).

The harmonic impedance Z_h of the upstream system viewed from injection point A is given by:

$$Z_h = \frac{V_h}{i_h}$$

The most current method uses a bank of reactive energy compensation capacitors as a source of harmonic currents.

Closing the bank causes a short-circuit, incurring a current step similar to a Dirac impulse whose spectrum is very rich in harmonics. From current

180 HARMONICS

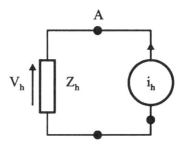

Fig. 8.2 Determining the harmonic impedance Z_h of the system seen from point A, starting from the injection of a harmonic current i_h.

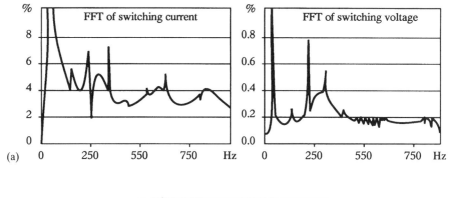

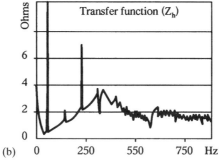

Fig. 8.3 Switching in a bank of capacitors on a 20 kV system. (a) Current and voltage spectra. (b) Impedance of the system viewed from the connection point of the bank.

and voltage readings at the terminals of the bank in a time window, including the transient thus created, signals and the harmonic impedance of the system viewed from the connection point of capacitors are obtained by the fast Fourier transform. Figures 8.3(a) and 8.3(b) give examples of current and voltage spectra obtained when closing a bank of capacitors in the 20 kV system, as well as its corresponding transfer function.

This measurement method is used very frequently to determine or check the harmonic impedance of systems experimentally. However, it cannot be used to analyse a future alteration of the operating schedule of a system. A digital method is used for predictive studies.

8.2.1.2 Computation of the harmonic impedance of systems

The impedances of systems can be determined by analytical computation. However, these computations soon become tedious as the size of the system

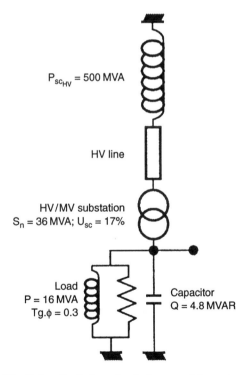

Fig. 8.4 Equivalent circuit diagram of the system.

increases. Some simulation models, such as Cymharmo developed by CYME, Inc. or Harmonique distributed by Electricité de France, make it possible to obtain the harmonic impedance of systems with up to several hundred nodes and meshes, with good accuracy.

For example, assuming that the effect of the length of the HV power supply line on the connection of a MV load producing harmonic currents I_h to a HV/MV source substation is to be examined. The equivalent system diagram is presented in Figure 8.4.

Figures 8.5(a), 8.5(b) and 8.5(c) present the harmonic impedances of the system at the injection point as obtained through simulation for respective line lengths of 50 km, 100 km and 150 km.

In general, the impedance of a system is formed of a succession of resonances and anti-resonances. These resonances are mainly due to the compensation capacitors and line capacitances. They exist at low frequencies in the case of very long lines or high installed compensation capacities. Their amplitudes can vary substantially depending on the load levels of systems.

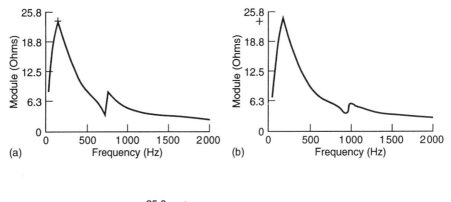

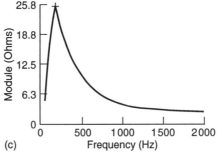

Fig. 8.5 Harmonic impedance of the MV system viewed from the connection point of the load for different HV line lengths: (a) 50 km; (b) 100 km; (c) 150 km.

In each case, the first resonance occurs at a frequency which depends on the length of the HV supply line. It varies between 165 Hz and 220 Hz. At this frequency, the amplitude remains practically constant, and is only limited by the equivalent resistance of the load connected at MV. A second resonance also appears in the neighbourhood of 1000 Hz, becoming gradually higher as the length of the line increases.

This simplified study reveals the undesirable risks of amplification of harmonic voltages around certain frequencies depending on the length of the HV line.

8.2.2 Harmonic current sources

Harmonic sources can be divided schematically into two categories depending on their origins: they can be intrinsic to systems or due to the nature of the connected loads.

8.2.2.1 Harmonics specific to systems

Although system lines do not introduce harmonics and often act as filters reducing distortions, other components contribute intrinsically to the deformation of the voltage wave.

Generators can be given as an example, which, despite optimized construction, do not deliver perfectly sinusoidal voltage (tooth-shaped harmonic), and mainly transformers which act as harmonic sources during their operation in saturated conditions.

8.2.2.2 System loads

Loads with nonlinear current–voltage characteristics connected to systems and fed by a practically pure voltage absorb non-sinusoidal voltages. These currents cross the Thevenin impedance of the system and generate voltage which is the more deformed the more intense the currents and the higher the impedance.

Among all these loads, a distinction can be drawn between two broad categories:

- random loads such as arc furnaces;
- loads supplied with power from devices including semiconductors, such as power converters and electrical domestic appliances.

Random loads

This category encompasses all units of equipment whose operating conditions are sufficiently obscure not to be able to determine, by a prior study, the spectrum of current absorbed by the load.

Arc furnaces are the main loads in this category. They are the more difficult to study as their power can reach extremely high values. It is nonetheless possible to determine experimentally empirical models of harmonic injections produced by these loads. For example, the arc furnace is modelled in harmonic conditions as shown below (example of an arc furnace model determined by experimentation).

The arc furnace is modelled according to the equivalent circuit diagram of Figure 8.6.

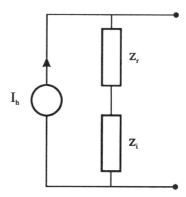

Fig. 8.6 Equivalent circuit diagram of an arc furnace.

Let S_n be the apparent power of the furnace, U_n the rated voltage between phases, and ϕ the argument defined by the power factor of the furnace $\cos\phi$:

$$Z = Z_r + jZ_i$$

$$Z_r = 1.2 \frac{U_n^2}{S_n \cos\phi} \qquad \frac{Z_i}{Z_r} = \tan\phi$$

The source of harmonic currents I_h is defined by:

Even order h: $I_h = \dfrac{S_n}{\sqrt{3}U_n} \cdot \dfrac{\left(0.15 + 3.5\exp(-0.4(h-2))\right)}{100}$

Odd order h: $I_h = \dfrac{S_n}{\sqrt{3}U_n} \cdot \dfrac{(0.15 + 7.5\exp(-0.45(h-3)))}{100}$

Loads supplied by semiconductor-based devices

The main reason for the success of semiconductors lies in the nonlinearity of their current–voltage characteristics. This special feature allows them to perform basic functions such as rectifying, or even dynamic power and velocity control. These harmonic current producing devices are classified into two categories as a function of the power of the loads they feed.

Power converters

This category comprises all equipment which perform the industrial system–load interface. Their operating modes are well known [4][5]. Their power levels are high and their applications manifold: electrical traction, electrolysis, induction, rolling-mill, etc.

All these applications require electronic switches (diodes, thyristors, etc.), as shown in the system of Figure 8.7.

With some approximation (perfect filtering, instant switching of thyristors, etc.), Figures 8.8(a) and 8.8(b) present the shape and spectrum of current absorbed by this device.

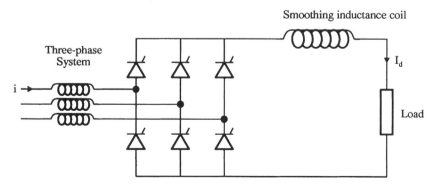

Fig. 8.7   Supply of a d.c. load via a Graetz three-phase rectifier bridge.

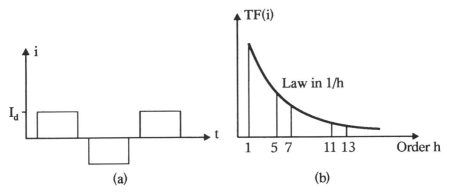

Fig. 8.8 Shape over time and spectrum of current absorbed by a six-phase rectifier.

An example of a model of a static converter

Let us take a static converter of apparent power S_n, with rated voltage between phases U_n and pulse order p (6 or 12).

Such a converter can be modelled by a source of harmonic currents i_h (h = order of the harmonic) such that:

Instant switching of thyristors:

$$I_h = \frac{S_n}{\sqrt{3}U_n h} \quad \text{with } h = pk \pm 1 \quad k = 1, 2, \ldots, n$$

Non-instant switching of thyristors:

$$I_h = \frac{S_n}{\sqrt{3}U_n h \left(h - \frac{5}{h}\right)^{1.2}} \quad \text{with } h = pk \pm 1 \quad k = 1, 2, \ldots, n$$

Electro-domestic loads

All electrical household appliances (television sets, video recorders, light dimmers, etc.), connected in millions of units to the low voltage distribution system, contribute to the greatest extent to harmonic pollution of systems. The first supply step of these appliances is formed of a diode bridge followed by capacitive filtering; other appliances are motorized via an electronic controller (in general, a triac switch); other appliances use discrete power regulation from a simple diode placed in series. They all generate substantial harmonic currents and contribute to a large extent to voltage wave distortion.

For example, Figure 8.9 presents the spectrum of current absorbed by a television receiver. To secure continuous voltage with low residual undulation, a capacitor is placed at the outlet of a diode bridge. It acts markedly on the shape of absorbed current and thus on the current distortion rate, whose values can exceed 100%.

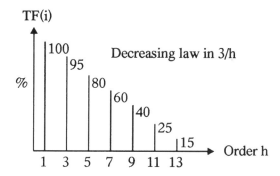

Fig. 8.9 Spectrum of current absorbed by a television receiver.

These electro-domestic loads produce high harmonic currents (law in $3/h$, compared with that of power converters which follow a law in $1/h$). In addition, despite the advantage of these appliances being of low unit power, their drawback is their substantial number distributed throughout the system. Owing to their extremely variable characteristics, they appear and disappear from the system at random. The great variation of these loads and lack of information on their numbers lead to their representation not with current sources of given amplitude, but with harmonic current sources which are statistically defined by density laws [6][7].

Taking a group of loads such as a group of television receivers, an apartment or even a distribution substation, it is shown that for each harmonic component h, laws (module and phase) representing the behaviour of this group can be drawn up. Of course, these laws depend on a number of parameters such as the season or time of day.

8.3 METHOD OF CALCULATION

As for any distribution calculation, a network containing N nodes is represented by N linear equations, or in matrix form:

$$I = YV \tag{8.3.1}$$

in which Y is the admittance matrix of the system; I designates the vector of harmonic current injections; V represents the vector of harmonic voltages.

Solving (8.3.1) makes it possible to determine:

- the impedances as a function of frequency at the different system nodes;
- the voltages and currents for each harmonic order at each node;
- the magnitude of the harmonic disturbance, measured by the voltage (or current) distortion rate TD at the different nodes. In the case of the voltage, this is expressed by.

$$TD_{\text{Voltage}} = 100 \cdot \frac{\left(\sum_{2}^{n} V_h^2\right)^{1/2}}{V_1} \%$$

in which n is the order of the maximum harmonic of the study, and V_1 is the fundamental voltage at rated frequency (50 Hz or 60 Hz);

- the voltage (and current) transfer coefficient as a function of frequency between two system nodes. This factor is used to analyse the effect of a particular component or of part of the system on harmonic propagation conditions (see section 8.4).

8.4 PROPAGATION OF HARMONICS IN SYSTEMS

In section 8.2 we showed that harmonic sources were represented by ideal current sources. These currents propagate in systems via lines and produce harmonic voltages at system nodes.

For a given harmonic order h, the voltage V_{h_B} generated in node B of a system by a harmonic current I_{h_A} injected in a node A (Figure 8.10) is expressed in the general form:

$$v_{h_B} = \frac{Z_{h_B} I_{h_A}}{\beta_{A \to B}}$$

in which: Z_{h_B} designates the harmonic impedance of order h of the upstream system viewed from node B; i_{h_A} represents harmonic current injected in A;

PROPAGATION OF HARMONICS IN SYSTEMS 189

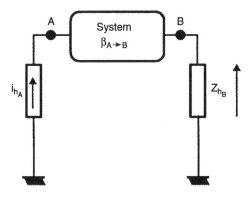

Fig. 8.10 Current transfer coefficient.

V_{h_A} represents harmonic voltage generated in B; $\beta_{A \to B}$ designates the current transfer coefficient of node A to node B defined by:

$$\beta_{A \to B} = \frac{I_{h_A}}{I_{h_B}}$$

This coefficient $\beta_{A \to B}$, which defines harmonic current propagation conditions in the system, depends on the relative impedances of connected loads and facilities. As a result, harmonic voltages can reach high values if injected currents meet high impedances or if currents are amplified during their transit in systems. The research into harmonic propagation phenomena aims to:

- determine the effect of each system component on harmonic injections (transfer coefficient study);
- define rules of harmonic propagation;
- draw up simplified charts equivalent to parts of systems.

To analyse these problems of harmonic propagation, a harmonic current or voltage is injected in a system node, and the resulting voltages between other nodes are examined on the basis of operating parameters of the system (topology, short-circuit power, load level, presence of reactive energy compensation capacitors).

8.4.1 Harmonic propagation in distribution systems

These systems have an arborescent structure. They consist of a HV/MV source substation feeding several MV outgoing lines which supply the MV/LV transformers to which LV loads are connected (Figure 8.11).

Reactive energy compensation is performed by shunt capacitors installed on the MV side of the HV/MV source substation or at the level of LV loads.

In harmonic conditions, the upstream HV supply system is basically inductive for frequencies which do not exceed 600 Hz. It is modelled by a pure inductance. It is a function of the short-circuit power at the rated frequency of the system at the HV connection point to the source substation and is defined by the relation:

$$L_{p_{scHV}} = \frac{U_{HV}^2}{\omega_0 P_{scHV}}$$

in which U_{HV} (kV) and P_{scHV} (MVA) designate, respectively, rated voltage between phases and short-circuit power of the upstream HV system, and ω_0 represents the angular frequency of the fundamental wave at rated frequency.

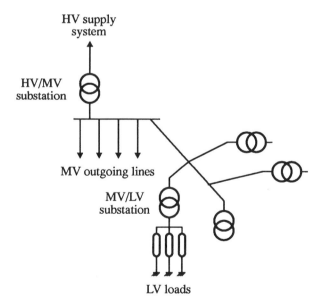

Fig. 8.11 Typical structure of distribution systems.

Short-circuit power can also be assessed at the level of busbars of the HV/MV source substation. For frequencies between 100 Hz and 600 Hz, the resistance, which creates the losses of the HV/MV transformer (see Appendix 8.A for the modelling of the transformer), is clearly higher in front of the leak inductance, so that:

$$P_{sc\,MV} = \frac{1}{\dfrac{1}{P_{sc\,HV}} + \dfrac{U_{sc}}{S_n}}$$

in which U_{sc} (%) and S_n (MVA) are, respectively, short-circuit voltage and apparent power of the HV/MV substation.

The equivalent inductance, viewed from the MV busbars, is then given by:

$$L_{P_{sc\,MV}} = \frac{U_{MV}^2}{\omega_0 P_{sc\,MV}}$$

Short-circuit power at the level of MV busbars is therefore greatly conditioned by the characteristics of the HV/MV transformer (Appendix 8.A). For example, short-circuit powers of French HV systems (63 kV and 90 kV) usually vary between 200 MVA and 1500 MVA, whereas those available at the level of MV busbars (15 kV, 20 kV, 33 kV) of source substations remain between 80 MVA and 200 MVA.

8.4.1.1 Propagation of harmonic currents from LV to HV

When the source of the disturbance is at LV, currents flow from LV to MV. This direction is opposite to the normal energy flow in the system, and the layout of impedances promotes the attenuation of harmonic currents. Figure 8.12 shows the system diagram in this configuration.

System without load and compensation

In this reference case, without loads and compensation, the propagation of currents depends only on the impedance values of the line and the transformation ratio of MV/LV substations.

The current first flows through the MV/LV substation and is subjected to constant attenuation, independent of frequency, equal to the transformation ratio of the substation. This current then propagates to the system via

192 HARMONICS

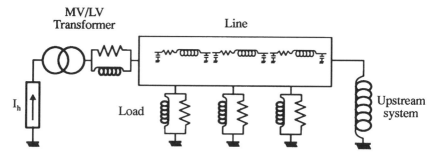

Fig. 8.12 Equivalent diagram of the system in harmonic conditions.

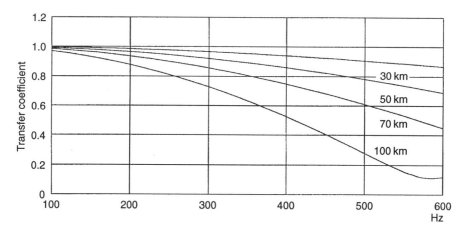

Fig. 8.13 Transfer coefficient of a 20 kV MV line as a function of frequency for different lengths.

the lines. Figure 8.13 presents the variations of the transfer coefficient of a 20 kV MV line as a function of frequency for different distances separating the injection from the observation point in the system. In the 100–600 Hz frequency range, it can be observed that, regardless of the length of the line, the transfer coefficient is lower than unity. The lines therefore 'amplify' harmonic currents.

It can nevertheless be shown that for lengths that do not exceed 30 km (typical length of distribution systems), the transfer coefficient remains between 1 and 0.90 in the 100–600 Hz frequency range.

As a conclusion, a harmonic current injection i_h, located at LV, produces a voltage of identical order to the secondary circuit of the HV/MV substation (Figure 8.14). This voltage is not very dependent on the location of emission in the system, and is only a function of the MV short-circuit power and thus varies practically in linear mode as a function of frequency.

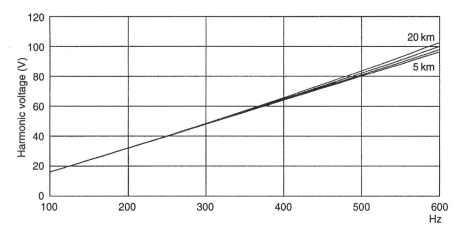

Fig. 8.14 Harmonic voltage at the level of the HV/MV source substation with a MV short-circuit power of 125 MVA, in the case of a 50 A LV current injection. Different distances separating the harmonic injection from the busbars are presented.

System with loads and compensation

The loads connected to the system consume part of the energy channelled by harmonic currents and therefore cause an additional reduction in harmonic voltages produced, increasing as a function of the load level of the system.

The reactive power capacitors installed in the HV/MV substation alter the harmonic impedance of the system. Their presence entails the creation of a filter plug formed by the combination of the upstream system short-circuit impedance and the capacitors. In the 100–600 Hz frequency range and with a certain degree of approximation (negligible influence of lines, low impedance of MV/LV transformers with respect to the LV loads they feed and very high inductances of loads with respect to that representing the upstream system), it is shown that the reduction coefficient provided by the loads can be expressed with a proper degree of approximation by the relation:

194 HARMONICS

$$\beta_{\text{Compensated load}} \approx \sqrt{\left(1 - h^2 \frac{Q}{P_{sc\,MV}}\right) + \left(hx \frac{S_n}{P_{sc\,MV}}\right)^2}$$

in which x designates the load level of the system; it is expressed as a percentage of rated power S_n of the HV/MV source substation and Q represents compensated reactive power ($Q = CU_{MV}^2 \omega_0$) installed at the MV busbars.

In the absence of compensation ($Q = 0$), the higher the load and the MV short-circuit power of the system, the greater the dampening effect of loads on harmonic voltage produced. At low frequencies, voltage remains identical to that secured in the reference case of Figure 8.14, and is subject to an additional reduction of 1.6 at 600 Hz for 40% load and 2.9 for 80% load.

The presence of compensation ($Q \neq 0$) gives rise to a parallel resonance frequency F_p defined by:

$$F_p \approx 50 \sqrt{\frac{P_{sc\,MV}}{Q}} \text{ Hz}$$

At this frequency, voltage produced at the level of busbars of the source substation is then at its maximum; it does not depend on compensated power but on the load level of the system (Figure 8.15). It is given by:

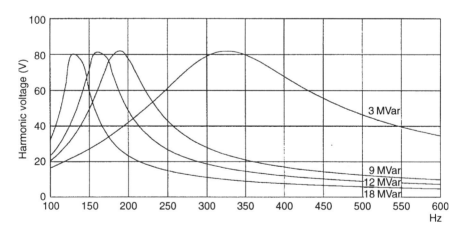

Fig. 8.15 Harmonic voltage in the secondary circuit of the HV/MV source substation with a 125 MVA MV short-circuit power, loaded at 40% of its rated capacity, in the case of a 50 A LV current injection. Four configurations are presented; a perfectly compensated system (3 MVar) and three over-compensated systems (9 MVar, 12 MVar and 18 MVar).

$$v_{h_{MV}}\big|_{F_p} \approx \frac{Ri_{h_{LV}}}{\beta_{MV/LV}} \approx \frac{U_{MV}^2 i_{h_{LV}}}{\beta_{MV/LV} x S_n}$$

From Figure 8.15 it can be seen that, for a given load level, the effect of over-compensation is to reduce the passband. However, the higher the volume of capacitors, the lower the resonance frequency, close to 150 Hz (danger of amplifying the harmonic of order 3 for a system at 50 Hz).

These same capacitors nonetheless promote voltage reduction when frequency reaches approximately $1.4F_p$ and should therefore be dimensioned to withstand over-voltages and over-intensities.

As a conclusion, a harmonic current injection at LV produces a voltage of the same order as in the secondary circuit of the HV/MV source substation. The lines do not act much on harmonic current propagation while their lengths do no exceed approximately 30 km. MV/LV substations reduce injected currents by a factor equal to their transformation ratio. The loads connected to the system provide an additional reduction, increasing in proportion to the load level of the system. The capacitors installed in the HV/MV source substation are at the root of a resonance between 100 Hz and 350 Hz, depending on the MV short-circuit power and the installed compensation volume.

8.4.1.2 Propagation of harmonic voltages from HV to LV

Figure 8.16 shows the diagram of the system in this configuration. The reduction depends practically on the ratio of impedances:

$$\frac{Z_{HV/MV} + Z_{line} + Z_{load}}{Z_{load}}$$

Owing to the orders of magnitudes of the line and the HV/MV transformer with respect to those of the loads, the reduction factor remains quite

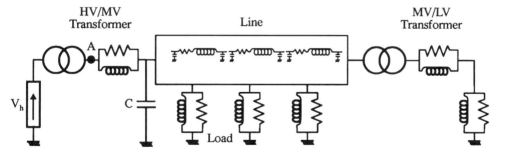

Fig. 8.16 Single-phase diagram equivalent to the system in harmonic conditions.

close to unity. In this configuration, harmonic currents circulate in the normal direction of power transfer in the system.

System without load and compensation

Without loads and compensation capacitors at MV, harmonic voltage propagation in the system depends basically on the impedance values of transformers and lines.

Voltage injected on the HV side is first reduced by a constant factor independent of frequency when flowing through the transformer of the source substation (transformation ratio). Therefore, everything happens as if we were injecting, directly at point A (Figure 8.16), a harmonic voltage with the amplitude:

$$V_{h_{MV}} = \frac{V_{h_{HV}}}{\beta_{HV/MV}}$$

This voltage propagates via the lines throughout the downstream system. In the 100–600 Hz frequency range, it is shown (as in the previous section) that line amplification remains between 1 and 0.95 for lines whose length does not exceed 30 km. This same voltage is then present practically in full at LV (within the transformation ratio of MV/LV distribution substation).

System with loads and compensation

In this system configuration, neglecting line effects, the equivalent diagram of the system is given in Figure 8.17.

Assuming that the impedance, as viewed from the primary circuit of the MV/LV substation, is very high by comparison with that of all the other loads and considering the loads as purely resistive in the 100–600 Hz frequency range, MV and LV harmonic voltages can be approximated by:

$$V_{h_{MV}} \approx \frac{V_{h_{HV}}}{\beta_{HV/MV}\sqrt{\left(1 - \frac{h^2 Q U_{sc\,HV/MV}}{S_n}\right)^2 + \left(h U_{sc\,HV/MV} x\right)^2}}$$

$$V_{h_{LV}} \approx \frac{V_{h_{HV}}}{\beta_{HV/MV}\beta_{MV/BV}\sqrt{1 + \left(h U_{sc\,MV/LV} x'\right)^2}} \approx \frac{V_{h_{HV}}}{\beta_{HV/MV}\beta_{MV/LV}}$$

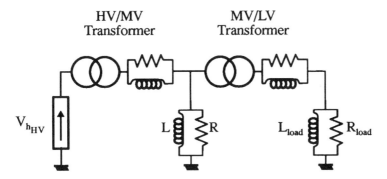

Fig. 8.17 Single-phase diagram equivalent to the loaded and non-compensated system in harmonic conditions.

in which: $U_{sc_{HV/MV}}$ and $U_{sc_{MV/LV}}$ represent the short-circuit voltages of the transformer of the HV/MV source substation and the MV/LV substation, respectively; and x and x' designate the aggregate load level of the system and the load level of the MV/LV substation considered, respectively.

In the absence of compensation ($Q = 0$), it is observed that the MV voltage reduction coefficient depends only on the load level of the system and the short-circuit voltage of the HV/MV transformer.

The reduction provided by an MV/LV substation does not depend much at all on the characteristics (U_{sc} and S_n) and the power of the load it supplies. LV harmonic voltage is therefore identical to MV within the transformation ratio.

The presence of capacitors installed in the HV/MV source substation changes the harmonic impedance of the system as viewed from HV and engenders a series resonance frequency F_s defined by:

$$F_s \approx 50 \sqrt{\frac{S_n}{U_{sc_{HV/LV}} Q}} \text{ Hz}$$

When compensation is adequate, resonance frequency F_s is around 250 Hz for a 70% load level and around 400 Hz if the system is loaded at 40% of its rated capacity. The reduction values are then minimal:

- for a 70% load level, 0.66, i.e. an amplification factor of 1.5;
- for a 40% load level, 0.47, i.e. an amplification factor of 2.1.

Beyond these resonance frequencies, harmonic voltages decrease more in comparison with the findings secured for a loaded non-compensated system. The reduction coefficient reaches the following values:

- for a 70% load level, its factor is 3 for a frequency of 500 Hz;
- for a 40% load level, the factor varies between 1 and 1.5 in the 500–600 Hz frequency range.

When the system is under-compensated, resonance frequencies are higher. The curves show that when frequency F_S is in the vicinity of 400 Hz, reduction is close to unity. When the system is over-compensated, resonance occurs around 150 Hz, and the minimum value of the reduction factor then becomes 0.45, which corresponds to an amplification of 2.2 for a 70% load level. Conditions are most favourable for the amplification of HV harmonic voltages when the system is over-compensated and its load level is low.

In sum, HV harmonic voltages propagate well to the lower voltage levels. Propagation conditions are independent of short-circuit power of the upstream system and the power of MV/LV substations connected downstream. They depend only on the load level of the system, and the value of compensation capacitors installed at MV. Without these capacitors, the reduction factor is between 1 and 2. By contrast, their presence causes amplification phenomena which can reach a factor 3. In the case of a high load level (70%), the frequency band concerned is based on 250 Hz and spreads to approximately 300 Hz. This range broadens for a low load level; it can cover the 50–1000 Hz band for a very low load level (20%).

8.4.2 Harmonic propagation in HV and EHV transmission systems

Two characteristics differentiate HV and EHV transmission systems from distribution systems, and necessitate an appropriate study. First, in contrast to the MV and LV systems with an arborescent topology, transmission systems are meshed.

Moreover, the dimensions of these systems, which are much larger than systems at lower voltage levels, are of the order of magnitude of the wavelengths of the harmonics propagating on these systems, thus risking the emergence of one or several resonances along the lines.

It is impossible to consider the overall system on account of these large dimensions. Consequently, it is necessary to limit the scope of study and model the remaining part of the system with boundary equivalents. These equivalents are represented by typical loads R, L calculated from active and reactive power transfers at the fundamental frequency at the boundary point considered.

In contrast to MV systems, there is no simple and satisfactory method to identify harmonic propagation in this type of system because the two special features mentioned above render their mathematical analysis extremely complex. To be able to take into account possible resonances along the lines, a new model with distributed constants more appropriate than the model in π has to be used.

Each infinitely small line section is represented by the following quadrupole of Figure 8.18.

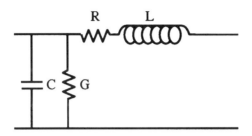

Fig. 8.18   Line element with distributed constants.

The electrical magnitudes, current and voltage meet the propagation equation (see Chapter 7, equations (7.3.13) and (7.3.14)), referred to as the 'telegram operators' equation:

$$\frac{\partial^2 v(x,t)}{\partial x^2} = LC\frac{\partial^2 v(x,t)}{\partial t^2} + (RC + LG)\frac{\partial v(x,t)}{\partial t} + RGv(x,t)$$

This model yields quadrupolar representations in T or π (Figure 8.19) which take into account the propagation phenomena of a line of length x.

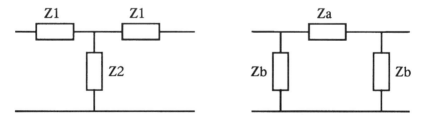

Fig. 8.19   Representation of a line element by quadripole in T or π form.

The impedance values of these quadrupoles are expressed in the form of hyperbolic functions of complex variables:

$$Z_1 = Z_c \frac{\text{ch}(nx)-1}{\text{sh}(nx)} \qquad Z_2 = Z_c \frac{1}{\text{sh}(nx)}$$

$$Z_b = Z_c \frac{\text{ch}(nx)+1}{\text{sh}(nx)} \qquad Z_a = Z_c \text{sh}(nx)$$

in which Z_c and n designate, respectively, the characteristic impedance and the propagation index of the line defined by:

$$Z_c^2 = \frac{R+jLw}{G+jCw} \quad \text{and} \quad n^2 = (R+jLw)(G+jCw)$$

In theory, this analytical approach makes it possible to determine the voltages and currents at all nodes. Nevertheless, this method is far too

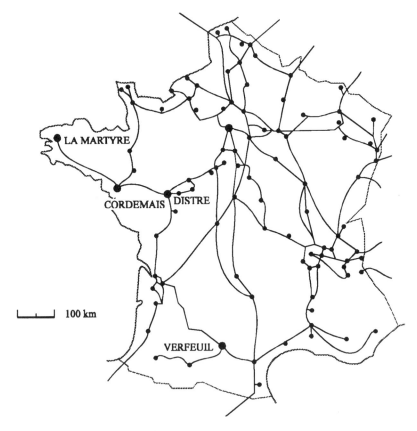

Fig. 8.20 French 400 kV transmission system.

complex from a practical point of view, because of the difficulty of the computations required.

Moreover, the presence of loops multiplies the number of equations needed to describe the system, thus necessitating long mathematical developments, even for relatively small systems. This explains why, today, simulation methods alone are used to analyse harmonic propagation conditions in meshed systems.

To illustrate these phenomena and the difficulty of interpretation of results, an example is given below of a simulation of an injection at the La Martyre (Brest) substation in Brittany (Figure 8.20). The resulting harmonic voltage in the 400 kV system is examined at the Cordemais (Nantes) substation, near Paris, at the Villejust substation, and at the Verfeuil substation (Toulouse) (Figure 8.21).

Although classical phenomena such as the presence of resonances, antiresonances and the reduction of disturbances with the distance ($V_{Cordenais} > V_{Villejust} > V_{Verfeuil}$) can be recognized in these graphs, it is yet quite difficult to

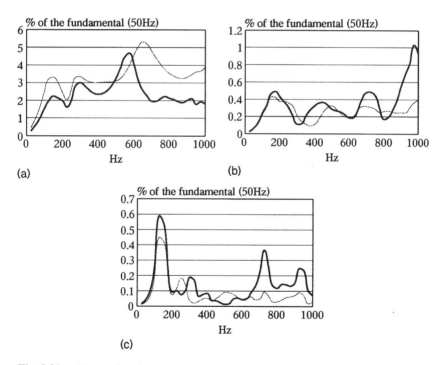

Fig. 8.21 Harmonic voltage, expressed as a percentage of the fundamental, at different points of the 400 kV ststem, resulting from a 100 A harmonic current injection at the La Martyre substation at (a) Cordemais, (b) Villejust and (c) Verfeuil where the Cordemais to Distre line was either in service (———) or disconnected (......).

give an accurate explanation for the variations in harmonic voltages as a function of the length of lines and frequency.

In conclusion, it should be recalled that since there are no analytical tools to study harmonic propagation in a large meshed grid, study of these phenomena remains a complex problem, from both the standpoints of its mathematical formulation and the interpretation of simulation results. These difficulties are basically due to the structure and characteristics of these systems. It can be observed that the effect of the increase in length of the line is to reveal a second resonance whose amplitude and frequency depend on the length of the HV line.

8.5 CONCLUSION

In the first part of this chapter, we presented the techniques and models permitting a description of the behaviour of systems under harmonic conditions. As a general rule, the choice of a model depends on the accuracy sought and the availability of the data. Moreover, it is important to remember that these modelling operations are valid only for frequencies not exceeding a few kilohertz.

In the second part, we briefly stated the principle of the method of calculation used in the majority of the software for simulating power systems under harmonic conditions. All the software now available on the market can be used to study the conditions of harmonic propagation on all types of networks with as many as several hundred nodes.

In the third part, we proposed an analysis of harmonic propagation conditions on distribution systems. We sought in particular to reveal the role of each element of the system in the harmonic injections and to establish simplified equivalent circuit diagrams for parts of the system.

Finally, in the last part of this chapter, we showed that research into harmonic propagation on meshed systems was still a complex problem on account of its mathematical formulation and the interpretation of the results. In contrast to distribution systems, for networks of this type there is no simple analytical method of depicting propagation phenomena, which means we must resort systematically to simulation tools.

APPENDIX 8.A REPRESENTATION OF SYSTEM COMPONENTS

All facilities forming systems can be represented in harmonic conditions using these simplified models. It should be borne in mind that these models are only valid for frequencies which do not exceed a few kilohertz.

Table 8.A.1 Values of the parameters for the different components (for indicative purposes)

Nature	Model	Description	Parameters		Remarks
R-L-C series branch		2 Nodes	R L C		
R-C parallel branch		2 Nodes	R C		
R-L-C series branch		2 Nodes	R L C	A B	R function of frequency
R-C parallel branch		2 Nodes	R C A	B	$R = R_0 \cdot (1 + A \cdot (\frac{f}{f_0})^B)$
Branch with N mutually coupled phases		2 Nodes	N [R] [L]		Poss. of giving zero and positive seq. parameters
Π model line		2 Nodes + earth	R L C	Length	
Line with distributed parameters		2 Nodes + earth	R L C	G Length	Poss. of taking the skin effect into account
N phases Π model line		2 Nodes + earth	[R] [L] [C]	Length	Poss. of taking the transportation into account
N phases line with distributed parameters		2 Nodes + earth	Positive and zero seq. parameters		Poss. of taking the transportation into account
Single-tuned filter		1 Node + earth	R L C		
Dampened single-tuned filter		1 Node + earth	R L C		
Stopfilter		1 Node + earth	R L C	R'	
Double-tuned filter		1 Node + earth	$r_1\ l_1$ r_2 r_3	C_1 C_2 C_3	
Type 1 load (R-L parallel)		1 Node + earth	R L	ou S ϕ	$R = U^2/(S \cdot \cos \phi)$ $L = U^2/(S \cdot \sin \phi \cdot \omega)$
Type 2 load		1 Node + earth	P ϕ		For remote control signals emission
Rotating machine		1 Node + earth	S_n	f $X_1 \%$	$R_p = f \cdot X_p$ $X_p = U_n^2/S_n \cdot X_i \%$
Single-phase transformer		2 Nodes + earth	V_1 R_1 L_1	V_2 R_2 L_2	Series model
Single-phase transformer		2 Nodes + earth	$U_{cc} \%$ S_n		Parallel model $R = 20 \cdot U_n^2/S_n$ $X = U_{cc} \% \cdot U_n^2/S_n$

8.A.1 Geveral system components

8.A.2 Transformers

Table 8.A.2 Representation of the three classes of transformers

Type	S_n MVA	U_{sc} %
MV/LV	0.05	4
	1.00	5
HV/MV	20	12
	36	17
EHV/HV	70	12

8.A.3 Lines

A line is defined by:

- its voltage level, U_n;
- its material (aluminium, almelec, copper, etc.);
- its cross-section (in/mm^2);
- its length (in/km).

The R, L and C parameters of the model with distributed constants are given in Table 8.A.3. The value of parameter G is zero.

Table 8.A.3 Parameters representing a line

U kV	15	20	63	90	150	225	400
R mΩ/km	460	460	160	130	120	60	30
L mH/km	1.33	1.33	1.21	1.21	1.33	1.26	1.05
C nF/km	20.0	20.0	20.0	19.0	17.2	17.8	21.6

8.A.4 Cables

A cable is modelled in the same way as a line, as in Table 8.A.4. The value of the different parameters alone changes. The material is aluminium, except for 400 kV for which copper is used. Units remain the same as above.

Table 8.A.4 Parameters representing a cable

U kV	0.4	5	15	20	30	63	90	150	225	400
R mΩ/km	260	260	260	160	160	50	50	40	30	15
L mH/km	0.27	0.30	0.32	0.35	0.32	0.32	0.32	0.32	0.32	0.32
C nF/km	660	660	660	660	660	660	660	660	660	660

APPENDIX 8.B DISTRIBUTION SYSTEM

The distribution system under study is a typical system representing an 'average' outgoing overhead line (Figure 8.B.1). It consists of a 36 MVA HV/MV substation feeding 7 MV outgoing lines. Only one outgoing line has been modelled. The others are represented by an equivalent load connected at MV to the busbars of the HV/MV source substation.

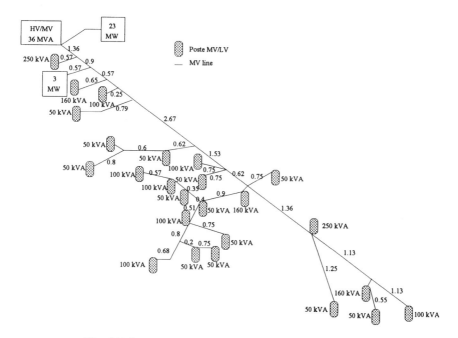

Fig. 8.B.1 Topology of the distribution system under study.

The capacity and length of the outgoing line modelled are 5.7 MVA and 50 km, respectively. It comprises two main lines approximately 25 km each.

One of these lines has been modelled in detail, whereas the other has been represented by an equivalent load of approximately 3 MVA connected to the separation point.

The modelled line is made of copper of cross-section $33\,mm^2$. It comprises 26 MV/LV transformers. These transformers are broken down below with respect to their capacity and by type of load fed:

Domestic loads (tg ϕ = 0.2)

8 transformers of 50 kVA
5 transformers of 100 kVA
2 transformers of 160 kVA
1 transformer of 250 kVA

Industrial loads (tg ϕ = 0.4)

5 transformers of 50 kVA
3 transformers of 100 kVA
1 transformer of 160 kVA
1 transformer of 250 kVA

HV (63 kV) short-circuit power is assumed to be equal to 310 MVA and reactive power compensation capacitors are installed at MV in the HV/MV source substation.

REFERENCES

1. Robert A. (1992) Guide for assessing the network harmonic impedance. Working group CC02, Cigre 36.05/Cired 2.
2. Bergeal J. and Moller L. (1983). Etude analytique de l'impédance spectrale d'un réseau: utilisation de méthodes numériques, Cired.
3. Lemoine M. (1977). Methods of measuring harmonic impedances. Intl. Conf. on Electricity Distribution, IEE Conf. Pub. 151.
4. Laborne H. (1989). ESE 6. Convertisseurs assistés par un réseau alternatif. 1– Concepts, critère de choix, dimensionnement, Eyrolles.
5. Laborne H. (1992) ESE 7. Convertisseurs assistés par un réseau alternatif. 2– Environnement, Eyrolles.
6. Deflandre T., Lachaume J. and Meunier M. (1993). Harmoniques sur les réseaux de distribution MT et BT – Niveaux actuels et futurs. Cired 2, 2.01.
7. EDF/DER (1989). Harmonique: Simulation des réseaux en régime harmonique et de la télécommande centralisée – Guide de l'utilisateur.

FURTHER READING

Arrillaga J., Bradley D.A. and Bodger P.S. (1985). *Power System Harmonics*, John Wiley and Sons.

Bieth J., Ollivier D., Lachaume J. and Meunier M. (1992). Perturbations harmoniques sur les réseaux français–Etat des lieux–Analyses–Tendances. Cigre 36.05, 205.

Escané J.M. (1987) ESE 5. Synthèse des circuits passifs et actifs–Filtres, Eyrolles.

9

DIGITAL REAL-TIME SIMULATION

9.1 INTRODUCTION

Study mode simulators are generally differentiated from real-time simulators. The former have no reason to respect real time and in general they serve for planning, operation planning or for designing new equipment, new controls or automatic controls. These study simulators are sometimes capable of running faster than real time, the aim being to obtain the most accurate possible result in the minimum time.

The need for real-time simulation of electric power systems arises basically for two major classes of simulators:

- simulators for training dispatchers in system management (which means simulators capable of testing new functions to be implemented in a control centre);
- simulators for testing the behaviour of equipment (controls, protections, automatic controls).

For these two classes of applications, the simulated phenomena will not necessarily be the same (long-term dynamics or electromechanical transients for the former, electromagnetic transients for the latter).

Real-time simulation aims to present the user or an equipment with phenomena and events as it would perceive them in reality. The real-time simulator is therefore subject to a stringent requirement, namely respecting the time imposed on it by the laws governing the phenomena concerned.

It should be remembered that there are analogue or hybrid (analogue and digital) simulators which have the same objective as digital simulators. When analog simulators were first developed, they were often the only way of simulating electrical power systems in real time (in particular for testing equipment when frequencies up to a few kHz must be represented). With the advances in information technology and the power of computers, there is now a strong trend towards the replacement of analogue simulators by fully digital simulators which are more easily configurable, and therefore less expensive to develop and maintain than analogue simulators. This chapter discusses real-time simulators, focusing on digital simulators only.

9.2 REAL-TIME SIMULATION FOR TRAINING DISPATCHERS

In the same way that aircraft pilots have to train on simulators, in particular to face emergency situations which they might never encounter in their professional careers, so the dispatching operators have to be trained to operate the system. This training of dispatchers has become current practice in electricity utilities, and is intended to train the operators on a representation of their own electrical power system. The training sessions aim to prepare them both for system management under normal conditions, and for highly disturbed and fortunately very rare situations on the actual system.

The simulator must therefore carefully represent the phenomena of the power system at each moment. It aims to produce a continuous authentic copy of the phenomena.

The art of simulation for training consists of taking the vision of the user to choose a clock and a reality to be simulated so that for the user everything happens as if he or she were observing the real phenomena. However, this identity between reality and simulation will be limited by the cost of the resources required. We shall see that the search for a compromise between quality and price can lead to accepting a plausible image of reality which is not an exact photograph.

An electrical power system is a system, which means a set of elements reacting with one another, whose behaviour can only be simulated in a monolithic manner given the speed of propagation of the phenomena. This particular feature allows us to consider that, in a simulator, the power system is a box which receives commands and supplies results. This box is the engine of the simulator. The specific problems of real-time simulation of electrical power systems lie chiefly in the engine of simulation. This engine solves all the equations governing the electrical system. Here we shall consider only the prime mover which is the first to suffer the constraints of time and on which the realism of a simulator is based.

In real time, time is an absolute standard. To respect it, a reference must be set: the clock. This clock must be selected to present the phenomena to the final user as they would perceive them in reality.

To comply with the constraints of real time, it will sometimes be necessary to introduce simplifications or to use tricks in digital computation.

9.2.1 Measurement of time

The use of digital calculation techniques imposes a discontinuous vision of time. A digital model makes it possible, when the state of a system at the time T is known, to calculate the state of the same system at the time $T + H$, and to do this within a time h. H is the time simulated or the interval of time, and h is the simulation time. What happens in reality between T and $T + H$ is not known.

Real time demands that the simulation time should be shorter than the simulated time. It does not in any way impose the sampling period H, the selection of which is of prime importance from two points of view: realism and difficulty, here synonymous with cost. Realism, since H is the minimum response time of the simulator to a request from the user, and it gives what one could call the degree of interactivity of the simulator. Difficulty, since the inequality $h < H$ can be difficult to respect for digital models. To give more freedom for the models, and thus indirectly reduce the production costs, the period H must be as large as possible.

The simplest way of choosing the period is *a priori* to consider only the viewpoint of the users. They alone can say what their needs are with regard to time. Is this method adequate? In the case of simulation in general, the answer is yes, since we are in control of time and modelling between T and $T + H$. This approach is found to be inadequate when we reach the technical limits of calculation: the calculating time h then becomes greater than H. In this case there are only two solutions: either we leave the dimension of real time, or we seek to reduce h.

This period H is the only time reference of the simulator which it is essential to respect to keep in real time. This constraint can be moderated in certain cases, by arguing the fact that simulation is not reality, and one can extend this argument to time itself if the user agrees. This transgression can take three forms: extended real-time, flexible real-time and frozen real-time.

9.2.1.1 Extended real-time

The principle of extended real-time involves not taking the standard second as the unit of time, but creating a fictitious unit of different duration. The referential system of time is simply changed. This is not always possible, in particular when a power system simulator is connected to an actual equipment having its own real-time clock. For example, let us take a simulator used to test protection equipment with a sampling period set by quartz crystal. Extended real-time has no real time apart from the name and the mechanisms, but it can be very useful if it is possible to expand the time without the user noticing this or being inconvenienced by it.

9.2.1.2 Flexible real-time

In flexible real-time, we shall take into account one special feature of power system calculation models: they solve differential-algebraic systems by iterative methods in which the number of iterations is a function of the disturbances affecting the power system. The calculating time h is therefore variable. H is selected as a function of a mean value of h and it is assumed that the disturbances causing a calculation time h greater than H are rare and are followed by relatively undisturbed periods during which h will be very small in relation to H. The principle is thus to allow a temporary time drift of the models which then become their own clock until their own time has caught up with real time.

9.2.1.3 Frozen real-time

The principle of suspended or frozen real-time is simple: when some models are too slow, the remainder of the simulator is frozen until their execution is completed.

The last two means presented above are only expedients and must be envisaged only in the very last resort. If one is obliged to have recourse to these, it is essential to provide monitoring of the drift.

Two elements emerge from this study of time: the definition of a period H and the fundamental hypothesis that everything which happens inside a calculation step h is seen from the exterior as being instantaneous. This will make it possible to define the power system, the simulated phenomena and the degree of realism of the simulator, that is the reality simulated.

9.2.2 Simulated reality

An electrical power system consists of an indissociable whole which goes from the generating resources to the consumers, passing through the lines, cables, transformers with monitoring and control devices: automatic controllers, protections and regulating devices. In general, we do not know nor can we simulate this ensemble absolutely correctly. It is therefore necessary to limit ourselves as to the type of phenomena simulated, the number of devices represented and the size of the system. These simplifications must not be made to the detriment of realism in simulation, which characterizes the quality of a simulator.

9.2.2.1 Limitation of the phenomena represented

The question then arises: what phenomena should be simulated? Of these, which should be represented as accurately as possible, and which can be represented approximately? To answer these questions our guide, contrary to the case of study mode simulation, is time. This applies from two aspects: first the total duration of the simulation, and second the simulation period H.

The total duration of simulation gives a very simple classification for electrical phenomena: those whose effects are felt during this period, and the others. This sometimes makes it possible to break down the problem in the case when the duration of simulation is low, just a few seconds.

In fact, even if all the characteristics of the power system and the phenomena affecting it are known, simulating these can prove to be too slow to respect real time.

Consideration of the period H allows another classification of electrical phenomena on two levels. The first level consists of phenomena whose time constant is greater than H, which must be modelled 'exactly', that is in the most accurate possible manner. The second level includes phenomena whose time constant is lower than H. Only the consequences of these are simulated. In other words, given the power system at the time t, we shall give its state at the time $t + H$ but shall not explain the way in which it changed from t to $t + H$, which is perfectly consistent with the discontinuous vision of time. The phenomena to be considered will be determined by the realism intended by the simulator.

In conclusion, exact simulation is provided for phenomena with a time constant greater than H, and the consequences of phenomena with a time constant lower than H are simulated.

9.2.2.2 Limitation on the representation of equipment

This limitation is divided into two categories: one imposed by the final use, and the other by the types of phenomena represented. The first is obvious: if the user can act on a device or see its operation, it must be modelled. The second is more difficult: it is a function of the realism sought in the simulation.

9.2.2.3 Limitation as to size

It is generally sufficient to simulate only what the user of the product can see, the rest of the power system being represented by an equivalent. A dispatching simulator will require a large system or network (several hundred buses), whilst a simulator for testing protection equipment can manage with a small network (around a dozen buses).

These last two limitations are not dictated so much by theoretical considerations as by constraints of cost and feasibility. Taking a large system or representing numerous automatic controls calls for the choice of more powerful and hence more expensive calculation resources, and it may be that there are no machines powerful enough to solve the problem.

9.2.3 The realism of simulation

We have defined the time and the reality of a simulator, and we now face the most important feature, the realism of the simulation. If indeed one can define the degree of realism as a divergence between what would happen in reality and what happens in the simulation, realism is achieved by the conjunction of the fineness and accuracy of the modelling. Fineness is a qualitative vision of realism whilst accuracy is a quantitative vision of it. The homogeneity of these two visions will give good realism to the ensemble. Here analysis of the final user's requirements only will make it possible to give the minimum fineness to the description of the power system, but this minimum is not sufficient; accuracy in simulation of electrical phenomena can imply finer modelling of the system. For example, if we wish to represent the phenomena of voltage collapse, not representing the automatic controllers reacting to the absence of voltage gives an unrealistic image of the general behaviour of the power system.

Enhancement of the fineness permits enhancement of the accuracy, and vice versa. The art of the modeller is to know how to satisfy the real needs of the user without giving in to perfectionism. Real-time simulation has the advantage of providing a good criterion for decision. In fact, whether it is for tests on equipment or training personnel, the final user will see the phenomena through a filter of known accuracy. In general it is not necessary to go further than that. This allows a field of validity to be defined, and a field of simulation to be deduced from it, thus obtaining the desired realism for which the results will be plausible, if not strictly correct.

9.2.3.1 Field of validity

This is the field in which simulation is correct. This field is delimited by the needs of the user. Its dimensions are relatively well defined and limited: the range of voltage, the range of frequency, the range of power of the system and the speed of variation of these variables. We shall model only the equipment likely to operate in this field, in which the digital algorithms must give correct results.

Determining this field is a choice just as vital as that of the period H. It governs the complexity of the algorithms to be used. Value analysis is a good guide here, establishing a link between the needs of the user and the constraints of the producer.

9.2.3.2 The field of simulation

Digital modelling does not have the same constraints as the physical approach, and to be certain of the results in the field of validity, modellers must define, under their own constraints, a field of operation of their models strictly including the field of validity: this is the field of simulation. Determining the field of simulation makes it possible to choose the algorithms.

The algorithms can be considered to give results in a field of simulation which the operation of the equipment will limit to the field of validity, thus giving homogeneous behaviour to the whole.

In comparison, the simulation of knowledge or study mode simulation takes place outside real time and its requirements are different. The criteria of choice in the fineness–accuracy dilemma are more tricky. The field of validity is in general no longer a given fact *a priori*.

In view of the above, it appears that each real-time power system simulator is a special case and the only way to tackle it is by an approach oriented towards the needs. Its needs, established by value analysis, will determine the choice of modelling.

9.2.4 Application

9.2.4.1 The Electricité de France training simulator

The example of the dispatcher training simulator developed in 1988 by Electricité de France can illustrate our subject. The training simulator has to represent the operation of a large electrical power system (over 200 substations and 300 lines) for a period of several hours with a sampling rate of a few seconds.

This simulator is intended for *in situ* training of dispatchers in the operation of a power system under normal conditions and disturbed conditions. This is a brief expression of the needs of the final user. What are the implications of this? From a general point of view, choosing a simulator reproducing the whole working environment of the dispatcher calls for the reproduction of their working tool: the dispatching or control centre. In this particular case, this reproduction is quite radical as the option chosen is to take the whole of the actual control tool and connect it to the power system simulator (the engine) as such. For the modeller, the users are more the dispatching centre and the instructor rather than the dispatchers since it is the former who determine the constraints of the models.

9.2.4.2 Constraints on simulation

Time constraints

At Electricité de France, the national or regional dispatching centres which the simulator must activate have a telemetry refresh rate of 10s. The period H defined above must therefore be 10s as a maximum. However, the simulator must be fully interactive and the maximum delay tolerated between the input of an action and its result has been set at 10s by the users. In view of the monolithic character of the simulation models, this delay expressed as a calculation step is 2; a command cannot be taken into account until the next calculation step. We are thus obliged to adopt a period of 5s.

Constraints on the phenomena represented

A simulation must cover a period of one shift for a dispatcher, that is several hours, making it necessary to model the long-term feedback of the power system finely. The simulator must therefore correctly take into account the frequency and voltage feedback from the power system following a change in the operating conditions of the generating and transmission system; the centralized controls such as frequency-power control will be represented.

The concept of disturbed conditions, apart from slow phenomena such as voltage collapse, has been extended at the request of the instructors to take into account short-circuits.

Constraints on devices represented

The dispatching centre has a view of the generating sets, the transmission structures, points of consumption and all the switchgear on the power system. Rotating loads are not specifically represented. The simulation of operation of the power system calls for consideration of the automatic controllers, regulating devices and protections of the system. It must be possible to take all these equipments out of service or make them fail at the request of the instructor.

Constraints concerning size

The power system represented must have the same size as the system monitored in reality by a dispatcher. Let us take, for example, 230 substations, 350 lines, 100 generating sets, 200 points of consumption and 3500 items of switchgear.

Field of validity

Taking into account the period of 5 s, the field of validity is divided into two subfields: the field of validity linked with phenomena having a time constant greater than 5 s, and that linked to other phenomena basically consisting of short-circuits.

For the former, the delimitation is simple: simulation must be valid up to the operation of the automatic controllers and protections. This corresponds to voltage between 0.4 and $1.5 U_n$ (nominal voltage), and frequency between 0.95 and $1.05 F_n$ (nominal frequency) for the power system. The conditions of the generating sets must respect their field of operation with regard to voltage, frequency, active and reactive power. Qualitatively, the accuracy must be consistent with that of telemetry, which is of the order of 1 MW, 1 MVar, 0.1 Hz and 1 kV.

For short-circuits, the field of validity is very restricted: only balanced dead short-circuits are processed, one at a time assumed to be cleared within the calculation step. The internal operation of the protection is not described, only the tele-indications and alarms generated by the protections shall be taken into account in the simulation models.

9.2.4.3 The solution adopted for the Electricité de France simulator

Representation of the power system

The system is represented in its detailed topology; all the switchable breaking devices are described. As only the balanced conditions are simulated, only the single-phase representation (positive sequence system) is provided. The system adopted has 200 substations, 350 lines, 100 generating sets and 150 points of consumption.

The operation of the short-circuit protections for lines and busbars, and overload protections for lines, cables and transformers is simulated. The automatic controls acting on loss of voltage, triggering circuits and transfers to standby voltage are also taken into account.

When it is a matter of simulating the national power system, the protections linked to the defence plan against lack of synchronism are also represented, but only approximately.

Choice of algorithms

For the phenomena with a time constant greater than 5 s, the field of validity is entirely covered by a long-term dynamic model (Chapter 5) which corresponds to the normal conditions of simulation. For high-speed phenomena, which here are due solely to short-circuits, the only objective of the models is to represent the operation of the protections and give the electrical status of the system once the short-circuit is cleared. We have therefore restricted ourselves to calculating the short-circuit currents and evaluating the stability of the generating sets by calculating the initial electrical impact and conducting only one approximate transient analysis stage. This is the disturbed condition.

Normal condition

In the normal condition, that is without a short-circuit, the model for calculating distribution is a slow dynamic model. This type of calculation is based on the hypothesis of linked rotors which assumes that all the generating sets connected to one power system are running at the same speed (common frequency). The model then solves all the classic equations of the active and reactive power balances at each node every 5 seconds and what is known as the rotating masses equation, which gives the modulus and phase plan of the voltages and the acceleration of the power system.

This very classic type of calculation has two special features here: the representation of the loads and the representation of the generating sets.

As the field of validity includes very low voltages ($0.2\,U_n$), the loads must have a realistic behaviour in the event of voltage collapse. With this in view, they are represented as an impedance seen via a transformer with a perfect tap changer.

This impedance is calculated at each step of the calculation as a function of the reference value given in P, Q.

Determining the long-term reactions of the power system implies a fine representation of the mechanical part of the generating sets. The modelling is in the form of differential equations. These equations are put into a form such that the mechanical power is written as follows:

$$P_m = \phi(\omega)$$

For lines, the system giving the active and reactive power throughputs is solved once per calculation step, that is for the link between nodes i and j (see Chapter 3, equations (3.3.2)).

For the loads connected at node i, the active and reactive power values are written as (following the notation of section 3.2.1):

$$P_i = G_i U_i^2$$
$$Q_i = H_i U_i^2$$

By writing the power balance at each node, we obtain the following system:

$$\sum_j P_{ij} + P_i = 0$$
$$\sum_j Q_{ij} + Q_i = 0$$

a system to which the dynamic equation of rotating masses must be added:

$$\sum_i P_{m_i}(\omega) - \sum_i P_{e_i} = \sum_i I\omega\gamma$$

In this expression: P_m = mechanical power; P_e = resistive electric power opposed by the system; I = moment of inertia of the rotating masses.

$$\gamma = \frac{d\omega}{dt}$$

is a differential equation integrated by the trapezoidal rule.

The fact of writing the rotating masses equation in the form $\gamma = d\omega/dt$ makes it possible to separate the solving of the differential equations for the mechanical models of the generating sets from the distribution calculation by assuming that the sum of electric power values is constant during one step of the calculation. One can therefore solve them with a smaller interval of time; in the example quoted their interval has been chosen at 0.5 s, which gives satisfactory realism.

With regard to the electrical part of the generating sets (alternator), the Potier representation is used.

Interaction between generating sets and the power system

The power system dynamics are chiefly due to the reactions of the generating sets faced with a change in the generating plan. The power system equations and the generating set equations should be solved at the same time. This is expensive in calculation time since the internal time constants of the generating sets require an interval of the order of 0.5 s. To avoid performing a calculation step for the slow dynamic model every 0.5 s, we have seen above that the coupling between the generating sets and the power system was expressed in a single equation: the equation of rotating masses. It is assumed that the electrical power demanded from the generating sets is constant during one calculation step of the system equations. The change from state T to state $(T + 5)$ s then takes place as follows:

- From T to $(T + 4.5)$ s, only the mechanical power of the generating sets changes. It is calculated by solving the differential equations which govern it by using the trapezoidal rule with an interval of 0.5 s.

- From $(T + 4.5)$ s to $(T + 5)$ s, the generating sets and the power system change at the same time. The power system equations, the rotating masses equation, the acceleration equation and the 'last' step of the generating set equations are solved simultaneously.

In short, one may say that it is not very accurate for 4.5 s but it is correct at the end of 5 s. This is in line with the discontinuous view of time.

Disturbed condition

In the event of a short-circuit, an approximate transient stability calculation step is performed (the generating sets are considered as electromotive forces, E, seen via their transient reactance, X'_d) which gives the value and the direction of the short-circuit currents and the initial acceleration of the power system by solving the following:

$$\begin{bmatrix} I_g \\ 0 \end{bmatrix} = Y \begin{bmatrix} V_g \\ V_c \end{bmatrix}$$

where V_g = voltages of the generating sets; V_c = voltages of the loads; Y = system admittance matrix; I_g = injection currents of the generators.

The simplified representation of the generating sets gives:

$$I_g = \frac{1}{jX'_d}(E - V_g)$$

The short-circuit is taken into account by modifying the matrix Y. The initial acceleration is calculated by solving the rotating masses equation and assuming that the mechanical power remains constant.

Architecture of real-time simulation software

This is derived from the discontinuous concept of time: everything which happens inside a calculation step is considered as simultaneous. This allows a calculation step to be cut into two parts: one part where the electrical state (moduli and phases of the voltages) is constant and only the topology of the power system varies; the other part where the electrical state varies whilst the topology is fixed.

Fixed electrical state

In this part, the consequences of the electrical state calculated at the previous step are analysed. It uses the following models:

- models of generators for which the change in mechanical power between T and $(T + 4.5)$ s is calculated as seen previously, determining the internal variables (stator current, excitation, etc.) and the possible operation of protections;

- models of automatic controllers reacting to the voltage (automatic controllers acting on loss of voltage, transfer to standby voltage, triggering circuits and others);
- a model taking into account the switching of breaking devices, which makes it possible to check whether the dispatcher has indeed followed the operational rules (analysis of switching).

These three models are 'issuers of commands': they just indicate which switching devices should be operated, without performing the commands. Then:

- Utilization of the model of topology changes the position of the breaking devices and determines the new nodal topology of the network.
- If there is no short-circuit (fed with power) on the system one can proceed to calculate the new electrical state.
- If there is a short-circuit on the system, the short-circuit currents are calculated and the protection devices which clear the fault are made to operate. The protections operate in a logical manner, this logic being based solely on the direction of the current passing through a breaking device. Breaking devices are switched and the topology program is used again. One can then proceed to calculate the new electrical state.

Fixed topology

Only the long-term dynamics model is used, which determines the new plan of moduli and phases of the voltages.

9.2.4.4 Typical use of the simulator

The modelling of the system has been defined on the basis of the user's requirements. We must now check that this modelling gives a simulation in accordance with the needs of the user. This *a posteriori* analysis takes time: the number of possible cases is almost unlimited. We shall restrict ourselves to a few aspects.

Interactivity

Modelling enables the user to operate all the switchgear, to modify the production–consumption–voltage space, to introduce failures or take the majority of the devices out of service. The response time to each action is 10s as a maximum. The limitations are located particularly around the failures which can be introduced. Actions can be freely linked together and each action is fully taken into account: the simulator does not include any concept of scenario. For example, the failure of a circuit breaker, introduced during the simulation, will result in a new sequence to clear a short-circuit if there is one. One may therefore say that interactivity is complete.

The case of a voltage collapse

The method of solution (complete Jacobian with calculation at each step) guarantees numerical convergence of the models up to voltages of around 1 kV. The representation of loads gives valid behaviour up to $0.15\,U_n$. Loss-of-voltage automatic controllers operate at $0.4\,U_n$. One can therefore say that simulation permits a faithful representation of the phenomenon of voltage collapse.

The case of a static loss of stability

This case is also known by the name of power transfer limit. The equation of the power carried on a line is a function which is limited if the voltages U_i and U_j are fixed. If the generating and consumption plan requires a throughput greater than this limit, the long-term dynamic model cannot find a mathematical solution. This is possible and occurs physically when part of the system loses synchronism. In view of the hypothesis of linked rotors, the phenomenon cannot be represented.

The case of a transient loss of stability following an event

The event giving rise to a transient loss of stability which naturally comes to mind is the short-circuit. In this case the approximate transient analysis used here evaluates the initial impact and can provide a plausible stability indicator which allows us to call upon the loss of synchronism protections, or not. This method of procedure is not strictly orthodox but is plausible and sufficient for operation training. The other event is the switching of a piece of switchgear. There, things are more complex since it is not possible, for reasons of calculation time, to carry out an approximate transient analysis at each switching operation. However, one can examine whether, in the whole of the switching operations performed, some create or interrupt a high throughput, and to perform transient analysis only in this case. Here we have reached a fundamental limit of the simulator and its behaviour can only be plausible if not strictly orthodox.

Resumption of service

The simulator cannot reproduce certain phenomena encountered during service resumption, such as ferroresonance and self-ignition. These phenomena correspond to very high-speed electromagnetic transients which cannot therefore be processed by a long-term dynamic model or a transient stability study mode model.

We thus see that, except for the resumption of service, modelling fulfils its role in a field of validity according to the requirements, but is limited in the representation of high-frequency phenomena. To extend its field, it would be necessary no longer to be limited by the hypothesis of linked rotors, and to move on to a true model of transient stability as described in Chapter 6, with the calculation time problems which that implies.

9.2.4.5 Return to time

The first function of a training simulator is to teach. For this reason, in addition to a strictly real-time operating mode, a slower mode and sometimes an accelerated mode are frequently requested.

The slower mode is useful for explaining certain types of behaviour of the electrical system and the simplest way of doing this is to use the concept of extended real-time. The standard clock is replaced by a fictitious clock at the level of the models. This has no influence at the level of the real-time tool: the values displayed will change more slowly. However, it can be advantageous for the slower timing to be accompanied by greater accuracy of simulation. One can also envisage an extreme solution which involves changing the modelling, which is not obvious from the viewpoint of connection of variables of state between two representations. A simple solution is to make better use of the existing representation by solving the mechanical equations of the generators and the power system equations simultaneously, and no longer once per calculation step.

The accelerated mode allows us to save time during periods of no particular interest or to rerun certain events quickly. This must be done without reducing the accuracy of simulation. It can therefore only have a very limited significance: the inequality $h < H$ is used in which h is the simulation time and H the time simulated, by making $H = h$ and eliminating the reference to standard time. This is possible only in a strictly diminished mode of use without connection to a real management tool, which itself is at real time.

9.2.5 Conclusion

Here we have mentioned only the real-time aspect of a simulation session, but the part outside real time must not be neglected: the preparation for and analysis of the session. For the final user at least, this part is just as important and is often the one governing the ease of use of a simulator.

9.3 REAL-TIME SIMULATION FOR TESTS ON EQUIPMENT

9.3.1 Why must equipment be tested?

To operate correctly, the electrical system requires numerous equipments: regulators, protections, monitoring and control devices. The safety of the system is dependent on their degree of dependability, so tests are therefore necessary:

- during the development phase, to optimize their design;
- before they are installed on the power system, to check their fitness to fulfil the specifications required;
- after their installation, in cases of dysfunction. It is a matter of analysing the causes of dysfunction and, if appropriate, determining the necessary corrective actions.

High-performance simulation tools are available to perform these tests, and obviously these must recreate the normal operating situations of the system and, above all, the disturbed situations which will allow us to measure the impact of the equipment studied. Phenomena of different types (electromagnetic and electromechanical transients) must be simulated as a function of the nature of the equipment to be tested. These tools must, in addition and above all, be able to produce and exchange data with the equipment *in real time* to respect the physical behaviour of the latter. This is a most important constraint which is not applicable to study mode simulators.

9.3.2 From analogue to digital

Up to the 1990s, only analogue tools could maintain real time at high frequency. There are numerous simulators of this type throughout the world (Canada, Japan, France, Italy, Sweden, Brazil, etc.). However, digital technology has many advantages compared to analogue: good representation of high frequency phenomena, greater ease of introducing new forms of modelling, faster and easier preparation for the test, and the reproducible nature of the simulations. This is why a trend towards everything digital has been noted in the mid 1990s. Up to then, this was limited by the level of performance of information technology resources. However, we have seen:

- the digitization of some equipment (rotary machines, control systems, etc.) in different analogue simulators which have thus become hybrids (Canada, France, ...);
- the creation of digital simulators operating outside real time (France) permitting the testing of equipment in an open-loop configuration, such as protections. These operate in two phases: the first relates to simulation of the power system outside real time, without the presence of the protection; the second relates to sending variables to the equipment tested in real time.

Since the beginning of the 1990s, the increase in power of the calculation facilities, in particular in the field of parallelism, has made it possible to produce fully digital tools for simulating power systems of sufficient size with good representation of high frequency phenomena (up to a few kHz). Achievements and projects in this field are increasingly numerous (Canada, France, Italy, etc.). We shall present the main characteristics of such simulators.

9.3.3 Digital simulators

A digital simulator chiefly comprises a simulation code for electrical power systems (or simulation engine), a computer, analogue/digital (A/D) and digital/analogue (D/A) converters for exchanging data with the equipment or equipment on test and a graphical user interface (GUI). The configuration is shown in Figure 9.1.

The phenomena to be simulated range from long-term dynamics to electromagnetic transients and the frequency band extends from 0 to 3 or

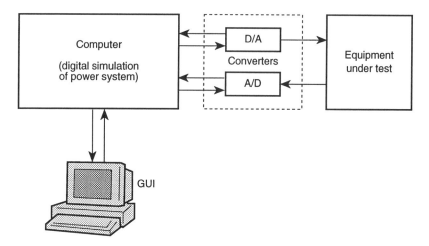

Fig. 9.1 General architecture of a digital simulator for testing equipment.

5 kHz. The integration time step adopted is generally around 50 μs. Ten rotating machines and 40 lines give an idea of the maximum size of the systems to be simulated (in 1996).

The main difficulty involves keeping to real time with such a small time step. The solution generally adopted is to utilize the parallelism inherent in solving the physical system. After describing the system to be solved, we shall present the possibilities offered by parallel information technology architecture for this type of simulator.

9.3.3.1 The equations of the electrical system

Transmission lines

The lines for which the propagation time is lower than the time interval adopted (50 μs) are represented by lumped R–L–C (i.e. standard resistance, inductance and capacitance in series) systems.

In the opposite case, the lines are modelled by distributed systems and called 'long lines'. The voltages and currents are then obtained by solving what is known as the 'telegram operators' equation. At each end and every instant, the electrical variables are dependent only on those calculated at the other end at the preceding instants. Long lines therefore make it possible to divide the power system to be simulated into **subsystems** which can be processed independently at each time (this the Bergeron method presented in Chapter 7).

The subsystems

The equations characterizing the operation of the different electrical elements forming a subsystem lead, after discretization, to a system of the type:

$$Y_{t-dt} U_t = S_{t-dt}$$

in which:
- t = current time;
- dt = interval of time;
- U_t = vector of the voltages at the nodes of the subsystem, calculated at the time t;
- S_{t-dt} = vector of the contributions of the electrical elements, in terms of currents, at the different nodes of the subsystem, calculated at the time $t - dt$;
- Y_{t-dt} = admittance matrix of the subsystem, calculated at the time $t - dt$ (known at the beginning of the interval of time t).

At each time t one should therefore:

- calculate the vector of the voltages U_t by solving the preceding system;
- calculate the vector of the contributions S_t of the admittance matrix Y_t which will be used in the next time step.

9.3.3.2 The key to success for real-time: parallelism

Solving the equations described above leads, taking into account a time step of 50 µs, to a calculating power of a few Gflops (billions of floating point operations per second) even for a system of relatively small size. At present they cannot be solved with one single processor; the key to success thus means utilizing the parallelism inherent in solving the physical problem. In fact, a distinction can be made between two levels of parallelism:

1st level: division into subsystems (Bergeron method)

As we saw previously, the subsystems delimited by long lines can be processed independently at each moment, that is in parallel.

2nd level: calculation of the contributions in terms of current per electrical element

For each electrical element (synchronous machine, line, transformer, etc.), the current flowing through it and its contribution to the admittance matrix can be calculated independently when the voltage at its terminals is known. The corresponding calculations can therefore be carried out in parallel for different electrical elements.

The parallel solution diagram for two subsystems is shown in Figure 9.2. The equations of subsystems 1 and 2 are solved in parallel at each time. For each of the subsystems, the vector of the voltages U is determined by solving the system $YU = S$. Then, in parallel, the current in each electrical element (or branch) belonging to the subsystem is calculated, the voltage at its terminals being known. These currents will serve to determine the value of the second member S which will be used at the following time.

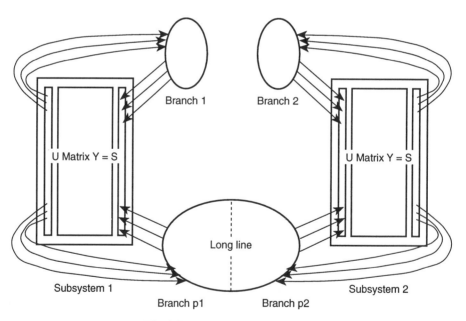

Fig. 9.2 Parallel solution diagram.

9.3.3.3 Possible information technology architectures

Needs to be met

Taking into account the two levels of parallelism (the second in particular) calls for extremely short communication times between processors. As a guide, a time of $5\mu s$ (that is approximately 10% of the time step), for exchanging 128 bytes between processors, appears to be entirely satisfactory.

The power desired for a processor depends on the complexity of the modelling used; 50 Mflops of peak power is a minimum value, and 200 Mflops or more would be far more satisfactory. The number of processors must be selected as a function of the size of the systems to be simulated

The flow of inputs and outputs must allow data to be exchanged very quickly with the equipment tested. A time of $10\mu s$ to transfer 128 bytes at the input or output will make it possible to test numerous equipments accurately (this time concerns only the inputs–outputs of the computer and does not include the times required by the converters, filters and amplifiers constituting the A/D and D/A interfaces.

Types of architecture now in use

The requirements stated above are very difficult to fulfil, and many teams throughout the world have chosen a dedicated architecture. The most significant example is that of the RTDS simulator (real time digital simulator) constructed by Manitoba HVDC (Canada) which uses DSPs (digital signal processors). Progress in the information technology industry now (1996) makes it possible to use standard parallel computers. EDF has thus demonstrated that such machines could satisfy the constraints stated and has selected a computer of the HP/Convex brand for its simulator.

9.3.4 Conclusion

For many years, electricity companies and manufacturers have been simulating electric power systems for testing equipment. Many changes have occurred, associated with the introduction of new equipment onto the system and the technical progress which, in particular, has made it possible to model electrical engineering equipment thanks to the use of electronic components. The changeover to the fully digital approach, made possible by rapid advances in the information technology industry, constitutes a challenge for the 1990s and several such systems have already been constructed throughout the world. It offers numerous advantages, both economic and technical: greater ease of use, better upgradability, and the reproducible nature of the simulations.

FURTHER READING

Bornard P., Erhard P. and Fauquembergue P. (1988). MORGAT: A data processing program for testing transmission line protective relays. *IEEE/PES*, 3(4), 1419–26, October.

Dommel H.W. (1969). Digital computer simulation of electromagnetic transients in single and multiphase networks. *IEEE*, **PAS-88**(4), 388–99, April.

Dommel H.W. (1992). *EMTP Theory Book*, Second edition, Microtran Power System Analysis Corporation, Vancouver, British Colombia, May.

Jeanbart C., Logeay Y. and Musart M. (1988). EDF's dispatcher training simulator, Cigré session, Paris.

Jerosolimski M., Descause D., Devaux O. *et al.* (1995). A real-time digital transient network analyser for testing equipment on a general purpose computer. Proceedings of the ICDS Conference, College Station, Texas, April.

Kezunovic M., Aganagic M., Skendzic V. *et al.* (1994). Transient computation for relay testing in real-time. *IEEE Trans. on Power Delivery*, **9**(3), July.

Logeay Y., Macrez J. and Meyer B. (1995). Training simulators for control center operators: EDF's past experience and projects for the future, IEEE Power Tech Conference, Stockholm, May.

Marti J.R. and Linares L.R. (1994). Real time EMTP-based transients simulation. *IEEE Trans. on Power Systems*, **9**(3), August.

Mercier P., Cagnon C., Tétreault M. and Toupin M. (1995). A real-time digital simulation of power systems at Hydro Quebec, Proceedings of the ICDS Conference, College Station, Texas, April.

Sekine Y., Takahashi K. and Sakaguchi T. (1993). Real-time simulation of power system dynamics, 11th Power System Computation Conference, Avignon.

Wierckx R.P., Giesbrecht W., Kuffel R. and Wang X. (1992). Validation of a fully digital real-time electromagnetical transients simulator for HVDC system and controls studies. Proceedings of the IEEE/NTUA Athens Power Tech. Conference, **2**, 751–59, September.

10

COMPUTING FACILITIES

10.1 INTRODUCTION

In the world of computer facilities used to simulate electrical power systems, digital computers are largely dominant (mainframes, workstations, personal computers of the PC type, etc.). This is partly due to the performances offered in calculation time and memory space, and also the flexibility in use for designing and operating the programs. Here we shall present two important aspects:

- the architecture of the information technology systems;
- the interaction between the user and these systems.

In these fields, progress is rapid and unceasing. The prospects opening up today permit effective and convenient use of calculation facilities whose performances place fewer restraints on study into electrical systems.

In the field of power system simulation, we saw in Chapter 9 all the constraints to be borne to achieve real time. For study simulation or simulation outside real time in general, the parameters of the problems are different. There, the user generally wishes to obtain the answer to a question as quickly as possible. In the past, effective calculating time was a limiting factor when undertaking simulation (the central processing unit (CPU) time). Today, in 1996, this limiting factor has diminished. The cost of the CPU is no longer, as on mainframes, calculated in time spent, and the hardware resources have become more available to the engineer.

On the other hand, the time spent by the engineer in translating a problem into simulation, in order to translate the results of the simulation back into the answer to the problem, has become a limiting factor. In fact, the engineer is not facing one question, but several. They must collect the data, conduct a dialogue with the software, run the simulations and analyse the results. The information technology architecture and the human–machine interfaces play a fundamental role here in increasing the productivity of the engineer.

10.2 INFORMATION TECHNOLOGY ARCHITECTURE

10.2.1 The problems

By information technology architecture, one should understand the whole of the hardware and software provided to make information technology resources available to a user, in this case a study engineer, to enable them to solve the problems they face.

In addition to the processing aspects, electrical engineering calculations and algorithms for system resolution or optimization, there are also the aspects of data description, modelling the power system and its behaviour, and the aspect of utilizing the results from the investigations. The working process of the study engineer is classically modelled as indicated in Figure 10.1.

Schematically, the beginning of a piece of study involves appropriating the available data and supplementing these data with information which is necessary but not available. Similarly, the end of the study work consists of presenting the results in order to draw the necessary conclusions from them, these conclusions being the object of the study.

The study or planning engineer, wavering between modifications to data or hypotheses, or the need to refine certain results, must then give free rein to their know-how. They find they have to handle large volumes of data or results, and concatenate numerous calculation codes. We thus see the infor-

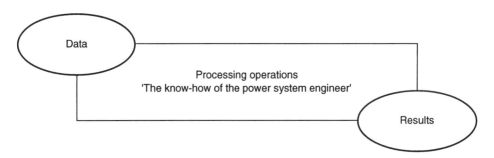

Fig. 10.1 Diagram of the power system analysis process.

mation technology architecture requirements for a study or planning analysis system emerging:

- For the architecture of the electrical system, they are:
  - the existence of a set of technical data references describing this system;
  - the connection of the study site to the information system of the company, to gain access to these data;
  - the ability of the user to handle large volumes of data at the workstation;
  - the possibility of exchanging data between study engineers or between companies.
- For the workstations, they are:
  - the availability of great calculating power to the final user;
  - the ability of the user to apply their know-how as well as is possible, via a high-performance GUI (graphical user interface).

10.2.2 The power system information system

Generally speaking, a mechanism for technical data management describing the electrical system must be designed for each electricity company. This is then known as the Power System Information System. This system must develop in accordance with the factors relating to the overall information system of the company. Extending beyond the particular requirements of simulation tools, associated fields handle the same data: system management, of course, plus invoicing, accounts, the construction of power system structures, and so on.

Care must be taken to implement suitable technologies and resources, so that the architecture of the information system (architecture of the data and the processing operations, hardware and software architecture) supports the strategies of the electricity company under the best conditions of efficiency, competitiveness, cost and safety.

The three major tasks of an information system are:

- to provide the users of the information system with data pertinent to the fulfilment of their duties;
- to bring the general organization of the data and the processing operations into line with the principles and rules of management of the technical function of the electricity company;

- to set up a hardware and software architecture which will be upgradable and easily configurable.

In this context the study engineer should be able to draw the data they need from a technical reference system. The data are then stored in databases. For example, at a central point, we find all the data for a study of the main transmission system of a company.

Extracts from these databases are then used to establish the initial data for study, as shown in Figure 10.2.

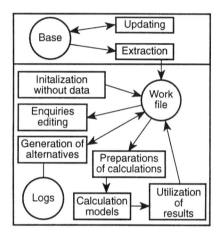

Fig. 10.2 Example of a planning system with technical references upstream of the investigations.

The nature of the data required for study work will of course vary with the type of study to be conducted.

Simulation of operation associated with the real-time concept calls for a very precise description of the structures to be indicated in detail. The greatest precision is necessary to describe the behaviour of the devices forming the power system. The view of the world to be simulated is that of one particular universe: certain events are taken into account in the study; others are not.

During planning, generally we use a somewhat less refined view of the power system, which is nevertheless sufficient to study its dimensioning. The view of the world to be simulated is much more uncertain. This is why a large number of hypotheses and random factors are taken into account. Some investigations are probabilistic, and provide results in statistical form. It is important to emphasize that data management, in the broad sense, represents a significant portion of the total study time.

INFORMATION TECHNOLOGY ARCHITECTURE

10.2.3 Computer techniques

First, a few facts. After the golden age of mainframes, the use of increasingly powerful workstations is a trend becoming widespread throughout the scientific community. The general development of scientific and engineering information technology is therefore continuing, accentuating the changeover to the following model:

- processing distributed on servers (data or calculation servers; this last case can be used through the use of parallel or vectorial computers), Unix workstations and microcomputers;
- data management on departmental servers or mainframe.

Communication is provided by long distance or local company networks, suited to the needs with regard to traffic density and transfer criteria. The communication networks offer or will offer very large transmission capacities and do not act as a brake on long distance exchanges.

Today, in 1996, a workstation comprises the following basic elements:

- central unit characterized by the type of processor and the clock speed;
- central memory with at least 64 megabytes;
- 1280×1024 colour video card and associated 19″ monitor;
- keyboard and mouse;
- mass storage of at least 1 gigabyte;
- devices such as DAT 2 Go reader, CD-ROM reader or recorder;
- the interfaces: serial, parallel, Ethernet, SCSI.

Obviously, these characteristics change rapidly as developments occur in the information technology market.

The upper limits in terms of disc or storage capacity are not indicated; the figures depend on the models. Very large values can be achieved from now on, and the costs involved are diminishing. Each manufacturer offers basic software with their machines and a catalogue of application software running on its platform. What the manufacturer offers generally includes:

- the basic operating system and a multiwindows facility;

- a network manager for connecting machines onto a single Ethernet network and distributing a system of files;
- development tools, compilers, linkage editors, etc.

However, this concept of the workstation must be supplemented by the concept of the server. In practice, the final user will work at a workstation, generally an ensemble of monitor, keyboard and mouse. In an applications context, there may be several users at a given moment, each having their own workstation. The problem then arises of the possible sharing of data processing resources between these users. The following can be distributed:

- central processing unit resources;
- storage capacities;
- multiwindows functions;
- the devices for saving, reproduction and communication;
- access to applications.

A server places data processing resources at the disposal of several users, and they are then shared via a local area network (LAN).

A local area network (LAN) makes it possible to link together the workstations, work positions or servers; LANs can in turn be interconnected via a wide area network (WAN) (Figure 10.3).

10.2.4 Prospects

Any development relating to information technology architecture should favour the open-ended design and modularity of study systems. It must be possible to accept the data needed for a new function without modifying the existing functions. To go further along this path, the introduction of a new code into the system must be envisaged from the outset by establishing rales which any new function must respect. The modularity of the design should thus make it easy to change an element of the system.

Like any other application, a study system should adapt to different variations in the information technology market during its service life. For this to be possible without fundamental modifications, only market standards are adopted, both for the hardware and the software.

With regard to hardware, workstations of the PC type or the Unix workstation type have appeared with ever-increasing capacities and with

INFORMATION TECHNOLOGY ARCHITECTURE 237

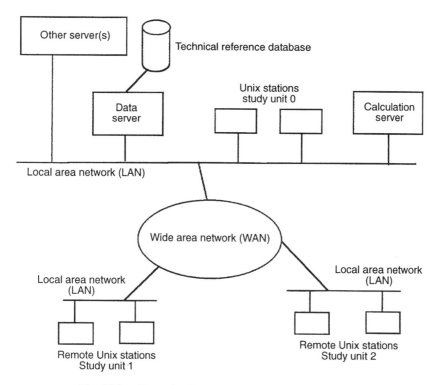

Fig. 10.3 Example of a network scheme at company level.

increasingly powerful graphic capabilities. Moreover, their costs are regularly diminishing.

With regard to software, Windows in the PC mode, and X-windows and Motif in the Unix world, permit the construction of user-friendly GUIs standardized from one application to another. Software tools now make the programming of such interfaces fairly fast. The appearance of relational database systems, with all their associated tools, has brought easy access to data for the users, upgradability and portability for applications, and speed of development for information specialists.

The maturity of object-oriented techniques in data processing (from design to languages or databases) already permits industrial developments which are proving to be more upgradable, and therefore less expensive to maintain than the older codes.

In the field of information networks, we have seen the Ethernet networks with the TCP/IP protocol providing easy communications in the Unix world. Software suppliers have produced extensions enabling them to be used via networks (particularly for databases). These developments lead to distributed information systems operating in client–server mode. At data distribution level, exchanges between companies are increasing, and networks such as Internet permit the exchange of data and software and give access to remote computers. Already many servers are found on the World Wide Web offering data exchanges or software as freeware. This free software, intended initially for universities, could partly take the place of industrial software.

The arrival of parallel machines, and the emergence of basic software permitting the construction of real distributed client–server architectures are doubtless the major developments which will have to be taken into account when producing the study systems of the future.

Tools for dynamic simulation of the electrical system lend themselves just as well to utilization on machines with parallel structure, containing from a few processors (CPUs) to a few dozen. Software of this type is likely to undergo several stages of parallelization, becoming more and more fine-grained.

Almost always several alternatives of the same case are examined; for example on the same network, the impact of a short-circuit appearing successively on each of the n lines potentially concerned will be analysed by dynamic stability software. Such simulations can be created for study into planning or in service, as a decision-making aid for the operators. This can be done by performing each elementary case in parallel on one of n parallel processors. Even if the final synthesis does not have the benefit of such a possibility of distribution, thanks to this single method of coarse-grained parallelization, it comes close to the theoretical maximum efficiency of computers (speed-up, which theoretically would want a saving in calculation time with a factor N for a machine with N processors).

However it is not applicable when only one case has to be studied, in particular in real-time simulation such as equipment testing or a training session for dispatchers (Chapter 9). Finer grains must then be sought. For example, we saw in Chapter 7 that Bergeron's equations decoupled the subsystems linked only by distributed elements: they no longer 'see' each other except by past terms. Electromagnetic transients can therefore be simulated on each subsystem on a different processor.

Finally, if all that does not suffice to take full advantage of the architecture of the computer used, we must seek, at the lowest level, to parallelize the basic algorithms. Different work has already been done throughout the world, in particular for load flow calculations.

To benefit from the progress opened up by parallel machines, it is now possible to take advantage of the emergence of real industrial machines equipped with all the necessary software tools. Similarly, new facilities of distributed architecture (groups of computers or workstations linked by high-throughput information technology networks) are also emerging from the channels now being explored.

Finally, among the prospects, one may only speculate at this stage about the future of the PC and Unix worlds. We see the power of PCs equalling that of Unix workstations. Common software standards are emerging. We see strategic alliances between manufacturers of Unix workstations and the producers of PC processors. The developments of power system simulators should at least guarantee portability from one world to the other.

10.3 THE GRAPHICAL USER INTERFACE

10.3.1 A few general principles

The term graphical user interface (GUI) covers all the solutions permitting a dialogue between the engineer and the study system. The purpose of the GUI is therefore to enable the user to use and obtain the best from the system.

Beyond the intrinsic interest of new calculation codes, a successful GUI is very often the means of making a user adopt a new system or new study methods. Similarly, the potential quality of study, determined by the characteristics of the processing operations, can be considered as an effective quality, determined by the GUI.

In the case of power system study, the major qualities expected of the GUI will be:

- performance: the faster the system can be used, the more scenarios can be envisaged in the study, and therefore the better the solutions found;
- user-friendliness: the easier the system is to use, the more the user will be tempted to get the best out of it, and therefore the better the quality of the study;
- transparency: the engineer responsible for study is not an information specialist; all the pure information technology aspects must be masked from them as far as possible;
- fitness for the task: the user must find it possible in their system to handle concepts with which they are familiar, and to find the electrical items they know and are studying;

240   COMPUTING FACILITIES

- customization: each user has a different concept of what a good study system is; giving them the means to customize their workstation is an incentive to better use of the system;

- simplicity: the simpler the GUI, the easier it is to learn, the faster and more fully it can be mastered; the presence of contextual and on-line help is a major advantage;

- standardization: faced with a plethora of systems for study, and production of documents, finding the same concepts at GUI levels is a necessity to avoid any rejection of a system.

The design of GUIs is part of the discipline known as ergonomics.

10.3.2   Development of GUIs

If we try to trace the history of user interfaces, naturally we follow the history of information technology. Without going back to the very first computers, it is useful to remember that in the early 1980s the natural user interface was on the one hand the tray of punched cards, and on the other hand the listing. The use of coloured cards and elastic bands then gave us the most effective user interface.

The appearance of passive screens and the conversational mode emptied the card trays in favour of file storage of the 'card image' type, fed by assisted acquisition. The user-friendliness of the dialogues lay in use of the function keys or lists by placing an 'S' in front of the lines to be selected. The concept of remanence between different sessions was then introduced. This allows the user to find their choices and parameters again. We speak of 'alphanumeric full screen' technology.

Two types of execution then appeared: deferred or batch processing, and immediate or interactive processing. Classically, the major calculation codes were batch processed, and the remainder were processed interactively. The tariff mechanisms of computer centres also influence the behaviour of the user: It is less expensive to carry out batch processing at night and interactive processing by day.

The first graphic reproductions on screen or on paper were then developed.

The arrival of workstations led to decentralization of the calculating power, and hence the possibility of designing interactive and graphic applications. The first applications on workstations provided the engineer with a study station where they could find things on a screen or keyboard and also a second screen, this time a graphical screen, where they could

THE GRAPHICAL USER INTERFACE   241

conduct a dialogue with the study system based on an image of the power system. The preparatory aspects of the calculations (preprocessors) and the reproduction operations (postprocessors) were then developed particularly well.

The emergence of technical solutions permitting multiwindows functions gave great flexibility in dialogue design. The user could have more than one screen at their workstation. New tools for dialogue appeared: scrolling menus, dialogue buttons, 'slide rules', etc.

These techniques really took off with the arrival of Unix/X-windows/Motif and MS-DOS/Windows. From now on, any application must have the *ad hoc* 'look and feel'. More than just a mode, the user finds an identical method of working among the applications they are led to use, and above all user-friendliness.

10.3.3   Prospects

The prospects should lie in three main directions:

- Complete control of the study system by the GUI: Instead of starting a calculation and asking for the results to be displayed, the request for a graphic display or the modification of a calculation parameter will generate the execution of the calculation, and more broadly, the different objects present on the screen will be linked to patterns of behaviour. The actions of the users on these objects or the interactions between objects will be the very essence of the simulation unfolding.

- The introduction of artificial intelligence and the concept of an aid to simulation will help the study engineer in their work, both during the course of the simulation and in the interpretation of the phenomena.

- The integration of simulation applications with other applications of the geographical or office equipment information systems type (spreadsheets, data processing, presentation assisted by multimedia computers) with concepts of dynamic links between objects from different applications.

Finally, the emergence of multimedia technology is already making it possible to transmit data from the machine to the user by hypertext (that is text, animated images and speech). Going beyond this, it is obvious that the interaction between the engineer and the machine should evolve at least as radically in the next ten years as it has changed since the time, not so long ago, of trays of punched cards.

FURTHER READING

Jerosolimski M., Descause D., Devaux O. *et al.* (1995). A real time digital transient network analyser for testing equipment on a general purpose computer, International Conference on Digital Power System Simulators, College Station, Texas, April.

Shlaer S. and Mellor S.J. (1988). *Object Oriented Systems Analysis: Modelling the World in Data*, Yourdon Press Computing Series.

11

NEW DEVELOPMENTS

11.1 THE COUPLING OF DIFFERENT TIME-SCALES

Dynamic phenomena on power systems are processed by three different types of models (long-term dynamics, stability and electromagnetic transients) and the operators feel that the less well justified the decoupling of the various types of phenomena, the more inconvenient this processing is.

Figure 11.1 illustrates the relative positions of the phenomena and the main types of models: it will be noted that the tools are not exactly adequate for the problems and that a problem such as the interaction between alternating current transmission systems and high voltage direct current systems, extending over practically all the time-scales, must be analysed by two or three different tools.

11.1.1 The physical origin of coupling

The increasing impossibility of separating the classes of time constants is fundamentally based on two trends:

- the desire to continue progressing in optimization of operations, which leads to operating a system in an ever more complex manner with ever narrower margins, simultaneously involving various phenomena close to the limits;
- technological progress in telecommunications and monitoring and control, which permits the construction of increasingly rapid monitoring and regulating systems.

Apart from the case of the alternating–direct current interaction mentioned above, we shall illustrate this observation by two examples. The maximum time for clearing a short-circuit at extremely high voltage (EHV)

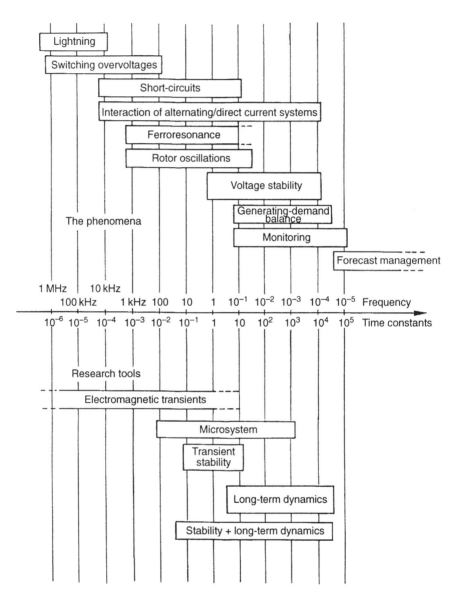

Fig. 11.1 The time-scales of phenomena and the main research tools currently available.

is constantly decreasing, since we seek to operate generating sets and transmission equipment to the utmost of their possibilities. In France today it is generally slightly shorter than a hundred milliseconds. In certain countries (the north west of the United States for example) it can be reduced to half this value, which requires the implementation of special techniques for circuit breakers, which in turn are controlled by very sophisticated ultra high-speed protection systems: any error in the calculation of this maximum time for clearing faults thus becomes very expensive, either in security if we mistakenly made it too long, or in needlessly expensive equipment if we made it too short. In the first thirty or forty milliseconds after an incident, the aperiodic components of the current due to the inductance of the lines and other electromagnetic phenomena in the generators (in particular the transformation electromotive forces mentioned in Chapter 6 in connection with Park's equations), all ignored in the stability models since they are incompatible with the hypothesis of the periodic nature of electrical variables, have a perceptible influence on the speed of the machines; for example, when a generating set is exposed to a short-circuit at its terminals, it brakes during the first moments under the effect of electromagnetic losses, before accelerating as any stability model predicts; this 'back-swing' phenomenon alone can cause an error of 30–40% in the results of the calculation of the time limit for fault clearance avoiding loss of synchronism. For correct analysis, it is then necessary to use a tool to study phenomena of the 'electromagnetic transient' class, which will give better results thanks to its more refined modelling. However, a very extensive portion of the electrical system has to be represented and its behaviour simulated for around 10 seconds; under these conditions the time and (cost) of calculation escalate enormously, which will limit the systematic investigation of the most stressful cases of operation.

The second example, which is much shorter, will illustrate the coupling between 'stability' phenomena and those arising from long-term dynamics. Automatic systems for centralized voltage regulation and generating control are becoming sufficiently rapid to interfere with the dynamic phenomena of the generating sets, particularly during widespread incidents. This will make obsolete the scrupulous classic practice of studying the different sequences of major incidents by alternating the use of stability programs and long-term dynamic programs.

11.1.2 The connection of the models

To meet these aspirations, bridges must be built between the ranges of tools.

We saw in Chapters 5 and 6 that the mathematical problems dealt with by the tools for research into stability and long-term dynamic phenomena were of the same type: in both cases it is a question of solving a problem based on algebraic equations of the power system and differential equations of the generating sets and regulating devices. The problem therefore arises only in the detailed formulation and method of solution; the first efforts therefore concentrated on the coupling of these two types of tools. We shall see below some of the benefits derived from this, and the gaps which still persist.

The link between stability and electromagnetic transients which require a different mathematical representation is more difficult: equations with partial derivatives on the one hand and algebraic-differential equations on the other. Indeed, it was possible to reduce the former to a real linear system, but by using the trapezoidal method with a very short integration interval (a few tens of microseconds) difficult to make compatible with the processing of the large time constants of generators. Today the linking of these two types of tools requires much further exploration.

A first step could be to extend the field of application of the stability programs by representing the power system in the form of differential equations (those of the 50 Hz or 60 Hz π unit) in place of algebraic equations. There would thus be a gain in accuracy in the low time constants range.

One could consider taking this type of representation further. In fact, if a line is modelled by a set of π units in series, instead of a single one, in the vicinity of the fundamental frequency this is reduced to integrating the telegraphy equation on a section of this line (Figure 11.2). We then obtain a first approximation of the propagation phenomena. This principle has been known for a long time since, in the form of a group of physical units, it gave rise in the 1950s to the first analogue tools for research into electromagnetic transients: transient network analysers. It must however be considered only as a way of extending stability models towards research into power systems with very large dimensions (with relatively low frequency propagation transients) and not as a means of bringing together stability and electromagnetic transient calculation programs. Besides the practical difficulties associated with the increase in the number of elementary bays

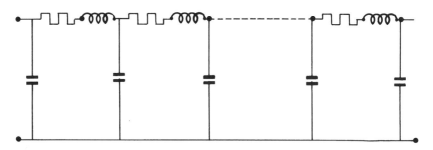

Fig. 11.2   Modelling a line by a set of ordinary differential equations.

and hence equations to be processed, this approach is based on equivalents at the fundamental frequency, and would therefore be unable to take into account phenomena of wave distortion or high frequency propagation. To do that, we must set down a single mathematical formulation covering the entire spectrum of high speed dynamic phenomena, and then find a suitable method of solution.

11.2 BRINGING TOGETHER STABILITY AND LONG-TERM DYNAMICS

Very recently, combined progress in computers and digital methods has permitted a major advance: the joining together of research into two classes of phenomena relating to transient stability and long-term dynamics in a single model. In this new broader class, the current and voltage waves are in a sinusoidal steady state, the generator rotors are free from any changes, and models of physical components are adopted to cover the entire frequency range concerned.

The period of observation of the electrical system can cover several tens of minutes, and the single new tool does not require more calculation time than its predecessors.

As we have seen, the problem raised is the integration of a large rigid algebraic-differential system (that is, one containing very different time constants), which is nonlinear, with discontinuities (caused either by events occurring on the power system, short-circuits, switching operations in substations, or by the regulating systems and processes).

This was solved by using a variable integration interval technique. The method used is based on two successive stages, prediction and correction. The values relating to the interval t_n are predicted on the basis of the values at t_{n-1} very simply by applying Taylor's formula in Nordsieck's formalism (Appendix 11.A).

This method leads to the formation of a nonlinear system which can be solved by the Newton–Raphson method, with LU factorization following an optimum order.

The integration algorithm obtained has the following characteristics:

- it is implicit;
- it simultaneously solves differential and algebraic equations;
- it has variable order and intervals.

It has the following properties:

- it is stable in orders 1 and 2, but gives an unstable response when the physical phenomenon simulated is unstable (this property is most interesting since we are seeking precisely to detect the instabilities of the system);

- it permits the detection of rapid excited modes and adjustment of the calculation interval to guarantee accuracy;

- thanks to special precautions, only major discontinuities (modification of the topology of the system, short-circuit, etc.) result in reformation of the Jacobian. Minor discontinuities (and frequent ones, for example when regulation reaches an end stop) are dealt with locally.

This method has been applied in the EUROSTAG software, designed and produced by cooperation between Electricité de France and the Belgian company Tractebel.

Moreover, EUROSTAG uses an automatic differentiation method to carry out a considerable number of function derivatives (Appendix 11.B). The method is efficient and can attain two apparently contradictory objectives:

- ease of insertion of the modelling in the form of block diagrams for which the expression of the derivatives is not explicitly supplied;

- reduced calculation time.

EUROSTAG unites the field of investigation operationally into transient stability as far as long-term dynamics phenomena, and is superior to each of the tools it replaces (Appendix 11.C).

11.3 NEW NEEDS, NEW RESPONSES

11.3.1 A need for ever increasing performance

We have seen that the simulation of slow dynamic phenomena called for precise representation of all the voltage levels, from the major transmission system to the automatic controllers for voltage regulation in distribution substations: it constitutes an in-depth view of the electrical system.

Research into stability phenomena is less demanding from this point of view: the modelling of the transmission system onto which the generators are connected is sufficient. However, electric machines play an essential role and do not lend themselves well to representation by equivalent units;

a broad spatial vision of the electrical system is therefore needed. The combination of these two constraints leads to research on networks in excess of a thousand nodes. In Europe, the development of cooperation and interconnection between the companies will further encourage this trend towards an increase in the size of the systems simulated.

When they study the behaviour of networks, engineers are aware of the response time of their computing system. However, it is more a factor of convenience (providing that it is not prohibitive), to be dealt with at the same level as two other points which we have already mentioned since, although they are of vital importance, too often underestimated in the old generations of tools, they lie outside the context of our discussion. These are firstly the management of the enormous volume of data required to simulate a case and its alternatives and on the other hand the ergonomics of the research tool, or rather the interaction between the user and computing tool (the concatenation of the different phases of research, and the utilization of the results must be via interactive tools fully exploiting the present graphic facilities such as multiwindows operation).

In two fields, however, a drastic reduction in simulation time is essential: real-time simulators (Chapter 9) and operation management support systems.

The latter require very short response times too. They are indispensable for managing the generating and transmission system as closely as possible to the cost–security optimum; at any time the control centre may have to take critical decisions.

This management has a random future (any line can be suddenly affected by a short-circuit, any generating set or major industrial load can be separated from the power system following a sudden fault) and the combination of possible responses by the operator is a non-polynomial problem. To explore the best solutions and prepare countermeasures, it is necessary to have very high-speed simulation tools, even though very effective methods are developed for the selection of contingencies.

11.3.2 The ways to progress

Meeting greater demands in terms of size and complexity of the systems studied, shortening response times sometimes as far as real time: these are the challenges arising today.

One way involves making use of new information technology architecture, particularly parallelism. In view of its importance, this point was dealt with separately in Chapter 10. Here we shall briefly present two other ways to research aiming to:

- develop new algorithms or refine their adaptation to specific features of power system problems;

- try to avoid simulation of the phenomena in terms of time but to appreciate the behaviour of the generating and transmission system by a direct quantitative evaluation. In this category we place in particular the tools for research into nonlinear instabilities such as those occurring on ferroresonant systems, the behaviour of which can be predicted with certainty by time-related simulations which still run the risk of allowing a bifurcation to a new dynamic state to escape.

11.3.2.1 Algorithms

In addition to research into new methods which could connect the models and renew the spectrum covered by the different tools, the field worthy of the greatest efforts today is that of solving very large algebraic-differential systems.

Continuous progress has been recorded, both with prediction-correction methods (by refining the adjustment of the Gear or Adams methods used in EUROSTAG, see Appendix 11.A), and with iterative methods of solving hollow linear systems (the GMRES, generalized minimal residual method, with preconditioning, permits excellent precision even with several thousand system state variables).

11.3.2.2 Direct methods

Direct methods for research into transient stability

When studying transient stability, the step-by-step simulation methods already described are the most well known and used. Their results are reliable and accurate, but they are expensive in computer time and do not provide synthetic results.

The principle of direct methods consists of comparing the energy stored during the fault with critical energy calculated beforehand. In theory this makes it possible to determine the stable or unstable state of the system, and the remaining stability margin. These synthetic methods, less greedy for computer time, permit rapid location of the potential risks. They have a potential field of application both in expansion planning to determine weak points in the system, and in operation management to show the operators the margins they have in relation to critical fault clearance times.

Unfortunately direct methods require a certain number of simplifications for their use, such as the absence of regulations and representation of the generators in the form of a constant electromotive force after a transient reactance. In addition, until recently, the results obtained were not very reliable for large power systems.

These disadvantages explain why these methods are seldom used. However, recent developments have revived the interest in this type of approach and applications on systems of considerable size can be envisaged.

As an example, one could mentioned two techniques developed by Electricité de France in conjunction with the University of Liège:

- the EEAC (extended equal area criterion) method, the principle of which involves dividing the machines into two groups (one called the critical group) and evaluating the critical times and margins of stability using a criterion based on the area;
- the DTTS (decision tree transient stability) method, which makes it possible to determine a stable or unstable state as a function of directly observable parameters (power generated by the machines, power flows on the lines, etc.). This result is obtained by prior statistical analysis of a large number of system states.

It should still be noted that direct methods will always require major simplifications to retain their effectiveness. Therefore they will never be able to achieve the precision obtained by step-by-step methods and will thus remain complementary to the latter, for example serving to choose the situations which must be studied accurately by time domain simulation tools.

Nonlinear oscillations in the field of electrical systems, and bifurcation theory

The electrical system may be subject to sustained oscillations whose frequency and amplitude characteristics are very far from the nominal values. In contrast to transients in linear systems, these abnormal operating conditions may well not die away and can persist until some modification in the system structure causes them to disappear.

Ferroresonance is one of these sustained nonlinear conditions. Its appearance, often unforeseeable and apparently random, is to be feared on account of its sometimes catastrophic consequences for the integrity of the system equipments (transformers, generators). The reputation of this phenomenon for complexity is linked to the fact that it cannot be easily reproduced by tests on site, and even less by digital simulations in relation to time. The great sensitivity of its appearance and form to the parameters of the system – the operating point and also the configuration of the network and the precise moment of switching operations – is one of the characteristic traits of ferroresonance and explains why the time domain simulation approach based on fully determined initial conditions, though often used in other fields, is inadequate here. It is unable to give the operator the only information of interest, namely the range of variation of the parameter delimiting zones at risk.

The theory of bifurcations of nonlinear dynamic systems is therefore a better framework for handling this problem, and using this allows the changes in the modulus of a dimensioning electrical variable (current, flux or potential difference) as a function of the variation in one or two 'sensitive' parameters of the electrical system (voltage level, damping resistance) to be followed up on a large scale.

These changes delimit ferroresonant zones in the space of the parameters, separated by folds, branches, discontinuities or other types of catastrophies, as can be seen in Figure 11.3.

This type of representation shows, much better than an isolated time domain simulation, the operating margins which lie between a given operating point and the nearest potentially dangerous ferroresonant condition.

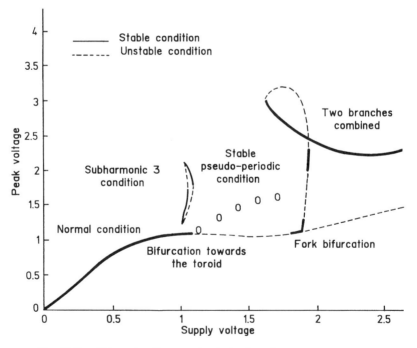

Fig. 11.3 Bifurcation diagram of a single-phase ferroresonant system.

Two problems still persist, however, limiting the use of these overall methods and indicating directions for progress. The first is of a theoretical type: the bifurcation diagrams can be used in the context of periodic non-autonomous or autonomous, continuous, nonlinear systems. The discontinuities such as modifications in topology imply a disruption in the structure of the system which means the entire bifurcation diagram must be recalculated at each switching operation on the system, which is expensive, even for a small system, in view of the considerable combinations associated with the various possible switching operations.

The second problem is linked with the size of the systems which can be handled at present using global methods, and the complexity and difficulty of interpreting bifurcation diagrams as soon as the number of system state variables and parameters increases.

As an example, most of the recent articles discussing ferroresonant bifurcations in the case of a transformer oscillating with a capacitor simplify the problem by ignoring the nonlinear coupling between phases existing in the saturable magnetic circuit of the majority of power transformers.

This gives us an idea of the ground which still has to be covered to deal with networks having several hundred nodes, using very fine modelling of all the nonlinear elements and great accuracy in the calculation of an equivalent of its linear part.

11.3.2.4 Simulation by modeller–solver–display devices

In a complete process of numerical simulation, three quite separate phases can be distinguished:

- the modelling which makes it possible to represent a certain physical reality by a series of mathematical equations, the complexity of which is a function of the area of observation examined;
- the creation of a computer model reflecting the mathematical formalism in numerical values;
- the use of the computer model including analysis of the results (plotting of curves, frequency analysis, the combination of elementary results, etc.) and also the construction of scenarios, sequential testing, etc.

Today these three phases are most often performed by specific models of the processes studied. In these models, the equations representing the process are frequently amalgamated with numerical solution schemes and they determine a fixed input structure. This overall integration has the advantage of generally obtaining good performances with regard to calculation time. However, the controlled performance of changes becomes difficult as soon as these models reach a certain size and one may also find subsets common to more than one model having different purposes.

A different approach consists of using modeller-solver-display devices:

- The modeller is a preferred tool for the first phase, thanks to the use of formalisms adopted in the field concerned (block diagrams, bond graphs, electrical equivalents, equations, etc.), and it can describe the components of the system and the way in which they are assembled. On the basis of this hierarchical description of the system to be simulated, it supplies the solver with a mathematical presentation in a suitable language, after having performed certain coherence checks on the type of interface variables between the components of the system, the dynamic description and the initialization scheme.

- The solver takes over and analyses the system structure, checks its regularity and conditioning and prepares one or more numerical solving schemes. It can also generate a simulator of the system studied, in a data processing language, which means a representation with a fixed structure of this system which is used by applying different scenarios to it, corresponding to different sets of parameters.

- The display device permits symbolic presentation during the simulation (curves, graphs, mimic diagrams, etc.) and postprocessing including, in particular, the processing of the signal, filtering, spectrum analysis and so on.

The advantage of an integrated modeller-solver-display device structure is that the postprocessor can find, on the basis of the main variables of the problem, a number of secondary variables such as the outputs of blocks in a block diagram, or the currents of branches in an electrical network resolved into nodal admittances. In fact, it can gain access to the internal structure of the system of equations representing the model. This means that fewer data need be stored and certain variables can be reconstituted on request without having to modify the equations of the model or restart a simulation.

This knowledge of the structure of the model also allows the creation of gateways at post-simulation level giving access to new functions such as harmonic analysis, identification and creation of dynamic equivalents, searching for natural modes, and optimization of regulating systems, etc.

Obviously, the more multi-disciplinary the modeller, the greater the advantage of the modeller-solver approach, which means it will adapt to different problems in research or engineering and that the solver will be of a general type capable of solving a broad range of mathematical systems.

In Appendix 11.D an application of this technique to the modelling of a hydroelectric equipment is described.

APPENDIX 11.A   EUROSTAG INTEGRATION METHODS

11.A.1   The problem

The modelling adopted for the simulation of electromechanical transients and long-term dynamics leads to the solution of a system in the form:

$$\dot{y}_j(t) f_j(y(t), t) \qquad \text{for } j = 1, \ldots, k$$

in which $y(t)$ is the state vector with k components:

$$y(t) = [y_1(t), \ldots y_j(t), \ldots y_k(t)]^T$$

11.A.2   The Nordsieck vector

At each moment in the simulation, for each component j of the state variables vector $y(t)$, its value $y_j(t)$ and its first q successive derivatives $y_j^{(q)}(t)$ are stored in a vector which has the following structure:

$$y_j(t) = \left[y_j(t), h\dot{y}_j(t), \frac{h^2}{2!} \ddot{y}_j(t), \ldots, \frac{h^q}{q!} y_j^{(q)}(t) \right]^T$$

in which h is the current calculation interval.

This Nordsieck vector is particularly well suited to changes in the calculation interval. Indeed, the new vector is obtained by multiplication by a diagonal matrix with

$$D = \begin{bmatrix} 1 & & & \\ & \alpha & & \\ & & \cdot & \\ & & & \alpha^q \end{bmatrix}$$

in which α is the ratio of two consecutive calculation intervals.

This formalism also facilitates changes of the order. In fact, a reduction of order is expressed by the elimination of the last line and an increase of order by the addition of a line, which is then easily deduced from the past of the simulation.

11.A.3 The Adams method of order r, implicit

The value y_{n+1} is obtained, the solution calculated at the moment t_{n+1} of the equation $\dot{y}(t) = f(y(t),t)$, by application of the formula:

$$y_{n+1} = y_n + \int_{t_n}^{t_{n+1}} P_n(t)dt$$

in which $P_n(t)$ is the interpolation polynomial of degree r verifying:

$$\begin{cases} P_{n(t_{n-i})} = f(y_{n-i}, t_{n-i}) & \text{for } i = 0, 1, \ldots, r-1 \\ P_{n(t_{n+1})} = f(y_{n+1}, t_{n+1}) \end{cases}$$

in which it is assumed that the values $y_n, \ldots y_{n-r+1}$ are known.

11.A.4 Method of retrograde differences of order r, implicit

These methods form the basis of Gear's algorithm.

The value y_{n+1} is obtained, being the solution calculated at moment t_{n+1} of the equation $\mathbf{y}(t) = f(\mathbf{y}(t),t)$, by application of the formula:

$$y_{n+1} = P_n(t_{n+1})$$

in which P_n is the interpolation polynomial of degree r verifying:

$$P_n(t_{n-i}) = y_{n-i} \quad \text{for } i = 0, 1, \ldots, r$$

and

$$P_n(t_{n+1}) = f(y_{n+1}, t_{n+1})$$

in which the values $y_n, \ldots, y_{n-r}$ are assumed to be known.

11.A.5 Choice of a method of integration

The two methods presented above are A-stable at orders 1 and 2 which are therefore the only ones used. This means that the simulation of stable phenomena with a larger calculation interval than the smallest time constraints of the system studied does not give rise to numerical divergence. They are therefore very suitable.

The method of retrograde differences in orders 1 and 2 is hyperstable. This means that an unstable physical phenomenon can appear to be stable under the effect of strong artificial damping. There are solutions for avoiding hyperstability but they are difficult to implement and lead to an increase in the calculation time.

The Adams method in order 2 is not hyperstable, whatever the calculation interval; the unstable nature of a physical phenomenon is retained. However, checking the error on the algebraic variables required for the quality of the simulation leads to short calculation intervals and hence high calculation times.

The use of a combination of the two methods has proven to be effective in the form of an algorithm with variable order and integration interval. The method of retrograde differences is applied to the algebraic variables; it is less sensitive to variations in algebraic equations than the Adams method. The Adams method is applied to differential variables, it ensures reliable detection of cases of instability.

These methods have been introduced successfully into the EUROSTAG program.

APPENDIX 11.B AUTOMATIC DIFFERENTIATION METHOD

11.B.1 Introduction

In many calculation programs we are led to calculate the values of a function which is not explicitly known but is represented by a data processing program. In addition, fairly frequently we wish to calculate the derivatives of such a function. Three methods are chiefly used to do this:

- numerical differentiation;
- symbolic differentiation;
- automatic differentiation.

Numerical differentiation is the simplest of these methods. It does not call for explicit knowledge of the program which represents the function. The derivatives are easily obtained by successively and imperceptibly varying each of the variables, whilst keeping the others unchanged. The latter are however affected by a numerical error which is greater or smaller depending on the form of the function. Moreover, and here is the main disadvantage, the calculation time is proportional to the time taken to obtain the function multiplied by the number of variables.

Symbolic differentiation [1–3] makes it possible to obtain a 'visual' representation of the derivative. Moreover, the formal calculation approach relieves the user of the task of writing a program. However, when the number of variables becomes considerable, the efficiency in calculation time tends to deteriorate. In addition, the program which represents the function must have an advantageous symbolic representation, which is not always the case.

Automatic differentiation, which was developed in the 1980s [4,5], makes it possible to obtain the derivatives of a function represented by a data processing program, whatever this may be. The derivatives are obtained accurately and above all in a calculation time proportional to the time for obtaining the function but unaffected by the number of variables. The ease of setting up is a substantial advantage of this method. There are even automatic differentiators [6–9] which generate the modified code for obtaining the derivatives of the function. The efficiency in calculation time of such a code and the memory space occupied are dependent on the form of the original program. Automatic differentiation does not however make it possible to obtain a symbolic representation of the derivatives, but only their values at a given point. Besides this, a great deal of memory space may be required.

The EUROSTAG program at present uses an automatic differentiation method to perform a large number of function derivative calculations. This efficient method has permitted the achievement of two apparently contradictory objectives:

- ease of integration of many modelling operations in the form of block diagrams in which the expression of the derivatives is not explicitly supplied;
- reduced calculation time.

11.B.2   The principle of the method

We consider a function $f: \mathbb{R}^n \to \mathbb{R}$ with n real variables $x_1, \ldots, x_n$ called independent variables. It is assumed that this function is represented by a data processing program. The aim is to calculate the exact derivatives of f at a given point $x \in \mathbb{R}^n$ in the shortest possible calculation time.

Let us take the example

$$f(x) = \frac{x_1 + \ln(x_2)}{x_3\sqrt{x_4}} - 2x_1$$

The value of this function can be obtained by means of seven intermediate variables which are also calculated by elementary functions (Table 11.B.1).

Table 11.B.1 Intermediate variables and the corresponding elementary functions (each line of the table corresponds to a program instruction)

Intermediate variables	Elementary functions
$x_5 = \ln(x_2)$	$x_5 = \varphi_5(x_2)$
$x_6 = x_1 + x_5$	$x_6 = \varphi_6(x_1, x_5)$
$x_7 = \sqrt{x_4}$	$x_7 = \varphi_7(x_4)$
$x_8 = x_3 x_7$	$x_8 = \varphi_8(x_3, x_7)$
$x_9 = x_6/x_8$	$x_9 = \varphi_9(x_6, x_8)$
$x_{10} = 2x_1$	$x_{10} = \varphi_{10}(x_1)$
$x_{11} = x_9 - x_{10}$	$x_{11} = \varphi_{11}(x_9, x_{10})$
$f = x_{11}$	

A representation in the form of a graph can also be used, each node representing an independent or intermediate variable, and each arc denoting the fact that a variable is dependent on another (Figure 11.B.1):

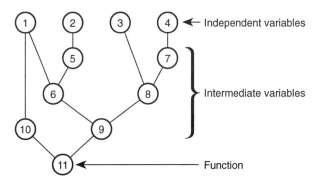

Fig. 11.B.1 Calculation graph.

The value of the function f is obtained by the composition of elementary functions. To calculate the derivatives of f, the rule of composite functions is applied. We then note that numerous products of elementary function derivatives are calculated in a redundant manner. The main idea of the method consists of cutting out a large number of these, saving them in dual variables.

The automatic differentiation algorithm consists of two phases, one progressive and the other regressive. The progressive phase corresponds to calculation of the function and derivatives of elementary functions; the graph in Figure 11.B.1 is crossed from 'top to bottom'. The regressive phase corresponds to the calculation of the dual variables and it allows the derivatives of function f to be obtained; the graph is crossed from 'bottom to top'. The method is of interest in that the ratio between the derivative calculation time and the function calculation time is independent of the number of independent variables.

11.B.3 Application to the simulation of electrical power systems: introduction into EUROSTAG

In the EUROSTAG program, the modelling of the electrical system leads to a large hollow algebraic-differential system. The integration algorithm used is an implicit prediction-correction method with variable interval which necessitates the calculation of many Jacobians.

One of the major advantages of EUROSTAG is the possibility offered to the user to enter graphically a certain number of modelling operations (regulation of generators, upstream of turbines, FACTS (flexible alternative current transmission systems) in the form of automatic block diagrams comprising elementary blocks (summing device, gain, integrator, time constant, dead band, relays, etc.). The price to be paid for this facility is an increase in the time to calculate the Jacobians, the functions and derivatives associated with these models not being known analytically. Automatic differentiation has been used to advantage.

In fact, the elementary blocks have only one output and correspond to a simple derivable function. The algorithm described above is applied using, as elementary functions, those associated with the blocks. The intermediate variables are the outputs of blocks. Automatic differentiation is therefore perfectly suited to modelling in the form of block diagrams.

11.B.4 Test results obtained with EUROSTAG

We have compared the EUROSTAG program integrating the automatic differentiation method with a version using numerical differentiation. The two give exactly the same numerical results, but with different calculation times. The relative saving increases with the number of variables. For example, for a system with 8000 variables, the saving achieved reached 80% for the Jacobian calculation time, which is 30% of the total calculation time.

APPENDIX 11.C APPLICATION OF A UNIQUE MODEL OF LONG-TERM DYNAMICS AND TRANSIENT STABILITY (EUROSTAG)

An electrical system is studied, comprising two subsystems A and B linked together by a long line (Figure. 11.C.1). Initially, the consumption of area B exactly balances its generation. An increase in load in this subsystem will introduce a power flow on the line linking the two zones. This will progressively reach its transmissible power limit.

To avoid total collapse, this line is opened and load shedding takes place in subsystem B. After stabilization, the two subsystems are reconnected.

Figure 11.C.2 shows the variation over a period of time of the angle of a machine in subsystem B. With EUROSTAG, only one simulation (a) will be necessary, whereas seven will be required if long-term dynamic software (b) is followed by transient stability software (c).

In addition, using such concatenation would not reveal the final loss of stability which does not occur following an event foreseeable by the research engineer; the latter will not therefore conduct the '4th simulation' of transient stability. The engineer will conclude that the electrical system stabilizes in accordance with the slow dynamics '3rd simulation', whereas in fact the network collapses.

Fig. 11.C.1 Electrical system studied, consisting of two subsystems A and B linked by a long line L.

262 NEW DEVELOPMENTS

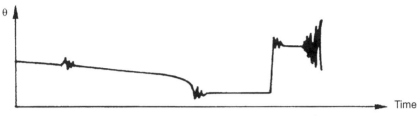

(a) Complete display of high-speed transients and long-term dynamics. Variable calculation interval (1 ms to 50 s)

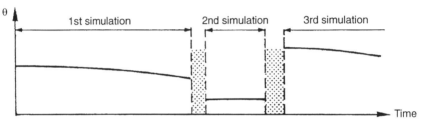

(b) Long-term dynamics software. Constant calculation interval (approx. 1 s)

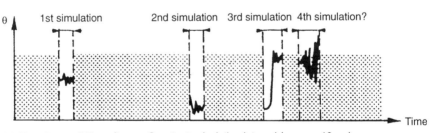

(c) Transient stability software. Constant calculation interval (approx. 10 ms)

Fig. 11.C.2 Change in the angle θ of a machine in subsystem B, calculated by EUROSTAG (a) or reconstituted by concatenation of simulations of the long-term dynamics type (b) and stability type (c).

APPENDIX 11.D   A MODELMAKER-SOLVER MODELLING APPLICATION

The objective is to construct a hydroelectric equipment model to perform dynamic operating investigations on the electrical system. More precisely, it concerns the hydraulic part including the turbines.

A library of elementary components was set up. By combining these components, a detailed description of any hydroelectric equipment can be created.

A modelmaker-solver-display device of the CASCD (computer aided control system design) class was used with block diagram language formalism, and using the Laplace transform.

The main blocks were thus defined as follows:

- the complete (flexible) Tube (long tube);

- the simplified (rigid) Tube (short tube);

- the Surge Chamber;

- the Orifice;

- the Pelton turbine;

- the Singular Loss.

As an example, Figure 11.D.1 shows the assembly of these elements to represent a part of the hydroelectric equipment.

Each element is in turn represented by a sub-block diagram comprising elementary components, like that shown in Figure 11.D.2 representing a penstock.

The parameters of Figure 11.D.2 are determined on the basis of equations of propagation under variable conditions in a penstock (Figure 11.D.3). In these representations: Q = flow; H = piezometric pressure in metres water gauge; L = length of the penstock.

By expressing the theorem of quantities of movement and the continuity equations, including the compressibility of the liquid and the elasticity of the walls of the duct, we obtain Allievi's differential equations for propagation under variable conditions:

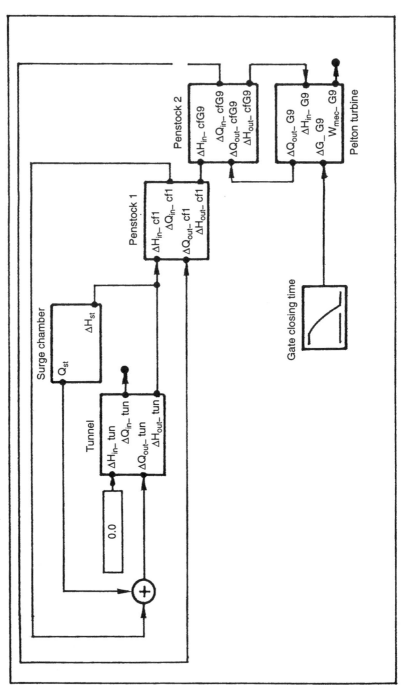

Fig. 11.D.1 Representation of a part of the hydroelectric scheme.

A MODELMAKER-SOLVER MODELLING APPLICATION

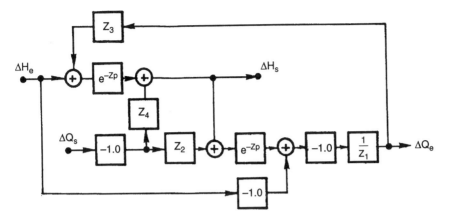

Fig. 11.D.2 Modelling a penstock in the form of sub-block diagrams.

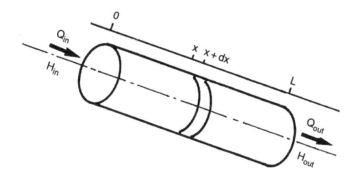

Fig. 11.D.3 Equations of propagation under variable conditions in a penstock.

$$\frac{\partial Q}{\partial t} + gp\frac{\partial H}{\partial x} = -\alpha|Q|Q$$

$$gp\frac{\partial H}{\partial t} + a^2\frac{\partial H}{\partial x} = 0$$

where: p = Laplace operator; t = time; x = abscissa; $H = H(x,t)$; $Q = Q(x,t)$.
The results are illustrated in Figure 11.D.4.

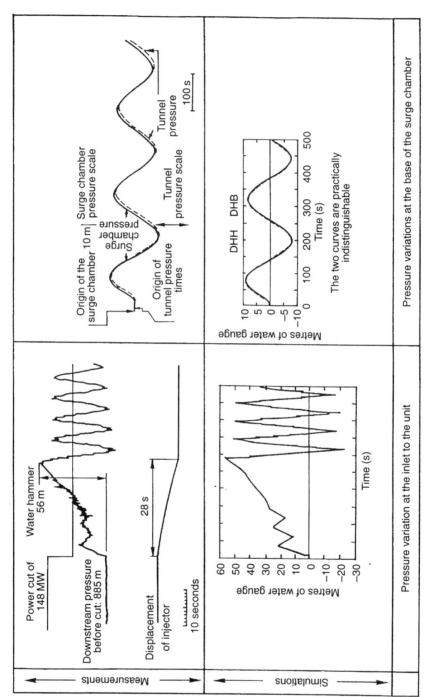

Fig. 11.D.4 Transients on tripping a hydroelectric unit: pressure head 885 m; power cut off 148 MW; flow cut off 19 m³s⁻¹; injector closing time 28 s.

REFERENCES

1. *VAX UNIX MACSYMA Reference Manual*, (1985). Version 11, Symbolics.

2. Char B.W., Geddes K.O., Gonnet G.H. *et al.* (1988). *Maple Reference Manual*, Symbolic Computation Group/Department of Computer Science, University of Waterloo, Ontario, Canada.

3. Hearn A.C. (1987). *REDUCE User's Manual*, Version 3.3, The Rand Corporation, Santa Monica, CA.

4. Gilbert J.C., Le Vey G. and Masse J. (1991). La différentiation automatique de fonctions représentées par des programmes (Automatic differentiation of functions represented by programs), INRIA research report No. 1557.

5. Griewank A. (1989). On automatic differentiation, in *Mathematical Programming: Recent Developments and Applications* (eds M. Iri and K. Tanabe) Kluwer Academic Publishers, pp. 83–108.

6. Hillstrom K.E. (1985). *User Guide for JAKEF*; Technical Memorandum ANL/MCS TM-16, Argonne National Laboratory, Argonne, Il.

7. Horwedel J.E. (1991). GRESS, a preprocessor for sensitivity studies on Fortran programs, in *Automatic Differentiation of Algorithms: Theory, Implementation and Application* (eds A. Griewank and G.F. Corliss), SIAM, Philadelphia.

8. Iri M. and Kubota K. (1990), *PADRE2, Version 1, User's Manual*; Research Memorandum RMI 90-01, Department of Mathematical Engineering and Instrumentation Physics, Faculty of Engineering, University of Tokyo, Hongo 7-3-1, Bunkyo-ku, Tokyo, Japan.

9. Griewank A., Juedes D. and Srinivasan J. (1991). ADOL-C, a package for the automatic differentiation of algorithms written in C/C++, Technical Report MCS-P180-1190, Argonne National Laboratory, Argonne, Il.

FURTHER READING

Alvarado F., Lasseter P. and Sanches J. (1983). Testing of trapezoidal integration with damping for the solution of power transients problems. IEEE/PES Summer Meeting, paper 83, pp. 364–7.

Astic J.Y., Bihain A. and Jerosolimski M. (1994). The mixed Adams–BDF variable step size algorithm to simulate transient and long-term phenomena in power systems. IEEE SM 93. *IEEE Transactions on Power Systems*, **9**(2), May.

Bergeron L. (1961). *Du coup de bélier en hydraulique au coup de foudre en électricité*, Edition Dunod, 1949. Translation: *Wave Hammer in Hydraulics and Wave Surges in Electricity* (translation committee sponsoring by ASME), Wiley New York.

Byrne A. Hindmarsh (1975). A polyalgorithm for the numerical solution of ordinary differential equations. *ACM Transactions on Mathematical Software*, **1**(1), 71–96.

Chaudhry M. Hanif (1986). *Applied Hydraulic Transients* (second edition), Van Nostrand Reinhold Company, New York.

Crouzeix M. and Mignot A.L. (1984). *Analyse Numérique des Équations Différentielles* (Numerical analysis of differential equations), Masson.

de Mello F.P. *et al.* (1992). Computations techniques for simulation of fast and slow dynamic effects in power systems, *IEEE Computer Applications in Power*, **5**(1), July.

EPRI EL 484 (1977). Power Systems Dynamic Analysis, Phase 1, July.

EPRI EL 4610 (1987). Extended Transient Midterm Stability Program, January.

Frankhauser H. *et al.* (1990). Advanced simulation techniques for the analysis of power system dynamics. *IEEE Computer Applications in Power*, **3**(4), October.

Gear C.W. (1971). *Numerical Initial Value Problems in Ordinary Differential Equations*, Prentice Hall, Englewood Cliffs, NJ.

Gear C.W. and Watanabe D.S. (1974). Stability and convergence of variable order multistep methods. *SIAM Journal of Numerical Analysis*, **11**(5).

Griewank A. and Corliss G.F. (eds) (1977). *Automatic Differentiation of Algorithms: Theory, Implementation and Application*, Society for Industrial and Applied Mathematics, Philadelphia.

Gross G. and Bergen A. (1977). A class of new multistep integration algorithms for the computation of power system dynamical response. *IEEE/PAS*, 293-306.

Hairer E., Norsett S.P. and Wanner G. (1987). *Solving Ordinary Differential Equations: Non-stiff Problems*, Springer Verlag.

IEEE Committee Report (1973). *Dynamic Models for Steam and Hydro Turbines in Power System Studies*. IEEE PES Winter Meeting, New York.

Jerosolimski M. and Levacher L. (1994). A new method for fast calculation of Jacobian matrices: automatic differentiation for power system simulation, IEEE PICA'93. *IEEE Transactions on Power Systems*, **9**(2), May.

Lorenz F. (1989). Acausal information bonds in bond graph models. IFAC Symposium AIPAC '89, Nancy, July.

Lorenz F. (1993). Discontinuities in bond graphs: what is required? Paper presented at the SCS conference in San Diego, (pre-print).

Marquet J.N. and Nicolas J. (1993). Modular modelling of a hydroelectric power plant; comparison between simulation and field tests records, IEEE, Athens, September.

Meyer B. and Stubbe M. (1993). EUROSTAG, a single tool for power system simulation. *Transmission and Distribution International*, March.

Pavella M. and Murthy P.G. (1993). *Transient Stability of Power Systems Theory and Practice*, J. Wiley.

Petzold L. *et al.* (1989). *Numerical Solution of Initial Value Problems in Differential Algebraic Equations*, North-Holland, Elsevier.

Sahlin Per (1988). MODSIM: A program for dynamical modelling and simulation of continuous systems. Proceedings of the 30th Annual Meeting of the Scandinavian Simulation Society, Helsinki, April.

Skelboe S. (1982). Time domain steady-state analysis of non-linear electrical systems. *Proceedings of the IEEE*, **70**(10), October.

Sowell E.F., Buhl W.F. and Nataf J.M. (1989). Object-oriented programming, equation-based submodels and system reduction in SPANK. Proceedings of the IBPSA Conference, Vancouver, May, 141–6.

Stangerup P. (1991). Requirements for a general purpose modelling and simulation language. Paper presented at the European Conference on Circuit Theory and Design, Copenhagen, September.

Stangerup P. (1991). Implementation of non-integer building blocks in general purpose simulation programs. Paper presented at the European Conference on Circuit Theory and Design, Copenhagen, September.

Stott B. (1977). Power systems step-by-step calculations. *IEEE 10th International Symposium on Circuits and Systems Proceedings*.

Stubbe M., Bihain A., Deuse J. and Baader J.C. (1989). STAG, a new unified software program for the study of the dynamic behaviour of electrical power systems. *IEEE Transactions on Power Systems*, **4**(1), February.

Systems Control Technology Inc. (1989). *Model C User's Guide*, Vols I and II, CAE Systems.

Vernotte J.F., Panciatici P., Meyer B. *et al.* (1995). High fidelity simulation of power system dynamics. *IEEE Computer Applications in Power*, **8**(1), January.

Working Group on Prime Mover and Energy Supply, Models for System Dynamic Performance Studies (1991). Hydraulic turbine and turbine control models for system dynamic studies, IEEE/PES Summer Meeting, San Diego, California, July.

Wozniak L., Collier F. and Foster J. (1991). Digital simulation of an impulse turbine: the Bradley Lake Project. *IEEE Transactions on Energy Conversion*, **6**(1), March.

APPENDIX A: THE MODELLING OF DIRECT CURRENT LINKS

The modelling of direct current links is complex. Here we shall briefly present the way in which direct current links are modelled, restricting ourselves to the case of symmetrical conditions, that is the system with positive components.

Before considering the modelling of these links, we shall give a brief reminder of their composition and mode of operation.

A.1 THE COMPOSITION OF DIRECT CURRENT LINKS

Direct current links comprise Graetz bridges installed in converter stations and connected by direct current links (lines or underground cables) (Figures A.1 and A.2).

In operation, direct current stations absorb the reactive power which has to be supplied by the compensation devices: capacitors, synchronous compensators or static compensators for reactive power, which are installed close to the stations.

Filters are also installed in these stations to absorb the harmonics created by switching the converters and to prevent these from propagating within the alternating current system.

A.2 MODE OF OPERATION AND CONTROL OF A STATION

A typical station is shown schematically in Figure A.3.

The current and voltage are regulated on the direct current side by influencing:

MODE OF OPERATION AND CONTROL OF A STATION 271

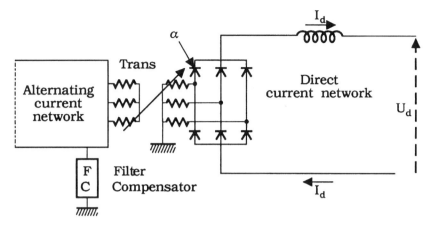

Fig. A.1 Converter station with one Graetz bridge.

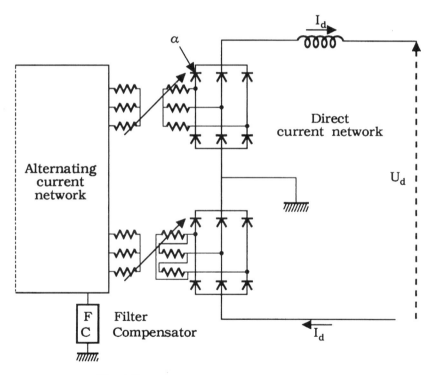

Fig. A.2 Converter station with two Graetz bridges.

272 THE MODELLING OF DIRECT CURRENT LINKS

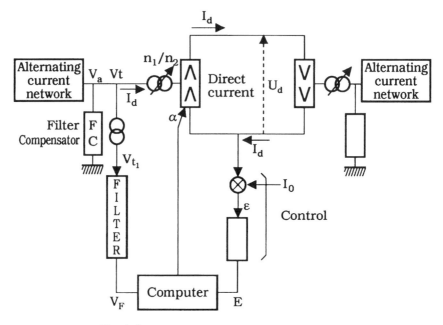

Fig. A.3 Schematic diagram of a converter station.

- the transformation ratio n_2 of the transformers supplying n_2/n_1 to the bridges: this time-delayed control is slow;
- the firing delay angle α of the rectifiers (thyristors) of the Graetz bridges: the control is almost instantaneous.

Two types of control are used at present: constant current control and constant power control.

In the first case, the direct current I_d is kept at a value equal to its reference value.

In the second case, it is the continuous power ($P_c = U_d I_d$) which is kept constant.

In general, the second type of control is superimposed on the first, and is therefore slower, in order not to interfere with it.

The direct current voltage U_d rectified is a function of:

V_a = alternating current voltage at the terminals of the converter;
I_d = direct current delivered by the converter;
α = firing angle of the valves;
$\left(\dfrac{n_1}{n_2}\right)_n$ = transformation ratio on tap n;

X_n = transformer leakage reactance for the transformation ratio n.

The corresponding relationship is of the following type:

$$U_d = f\left(V_a, I_d, \alpha, \left(\frac{n_2}{n_1}\right)_n, X_n\right)$$

A.3 CHARACTERISTICS OF A DIRECT CURRENT STATION WITH $U_d = f(I_d)$

For a given alternating current voltage V_a at the terminals of the link, the characteristic defining the voltage U_d on the direct current side is a function of the direct current I_d delivered, with the shape of the curve given in Figure A.4.

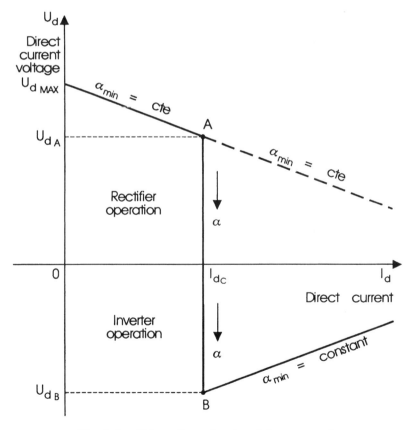

Fig. A.4 Voltage U_d as a function of direct current I_d.

When operating as a converter, the transformation ratio is assumed to be constant:

$$\left(\frac{n_1}{n_2}\right) = \text{constant}$$

For a constant firing angle α ($\alpha = \alpha$ min of the order of 15°), the direct current voltage U_d decreases when the current I_d increases (for example when the resistance of the external circuit decreases, thus increasing its load). This reduction is caused in particular by the voltage drops in the transformer leakage impedances.

The control can be used to keep the direct current I_d equal to a reference value I_{dc} whatever the value of the direct current voltage imposed by the external circuit onto which the converter station delivers power, whilst this voltage remains between two values U_dA and U_dB of the direct current voltage.

For this purpose, it is sufficient to change the firing angle of the rectifiers to compensate for the variations of the voltage U_d.

The vertical part AB of the station characteristic is obtained.

We note that when the voltage is reversed, the station changes from rectifier operation to inverter operation. To distinguish between these two types of operation, the firing angles are generally differentiated and called α and γ respectively.

A.4 MODE OF OPERATION OF TWO STATIONS CONNECTED BY A SINGLE DIRECT CURRENT LINK

If two stations are connected by a direct current link, one operating as a rectifier and the other as an inverter, and if the reference values of their settings are slightly different (I_{d1} and I_{d2}), their operating diagram is then as shown in Figure A.5.

The operating condition of the whole is located at the intersection of the two characteristics (at point M). The direct current voltage on the link is then the voltage U_{dn} and the direct current which flows on it is the current I_{d1}.

It is generally observed then that the direct current voltage is fixed by one of the stations, generally the inverter, and the current is imposed by the other station, generally the rectifier.

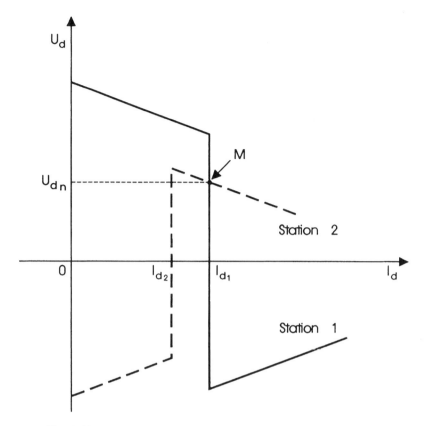

Fig. A.5 Operating characteristic of a link with two converter stations.

A.5 THE MODELLING OF DIRECT CURRENT LINKS

According to the studies undertaken and the desired degree of accuracy, direct current links can be represented with a greater or lesser degree of detail.

a) In models used for calculations in which, for steady state conditions, only the active power flows are of interest, since the power flows on the link are quickly and easily controlled by rectifying a greater or lesser portion of the voltage wave by acting on the thyristor gates, it will be assumed that the power carried in the link is set and constant. This power is by definition equal to the desired reference value.

In these models, a direct current link is then simply represented by two injections of power, equal in modulus and of opposite signs, at the two end nodes of the link.

b) For the alternating current distribution calculations, that is determining the voltage map of the system and the active and reactive power flows in all its elements, the representation of the direct current links is slightly more complex.

The equations which govern the operation of the network are divided into three systems:

System 1 — equations relating to the alternating current network.
System 2 — equations relating to the direct current network.
System 3 — equations relating to the converters.
Each system is solved separately:

- In system 1, the direct current network is not taken into account. It is simply represented by equivalent active and reactive loads connected to the alternating nodes where the converters are connected.

- System 2 is defined by Kirchhoff's equations relating to the direct current network: $I_d = Y_d U_d$.

- System 3 includes as many independent equations as there are converters. These equations take the following form:

$$U_d = f\left(V_a, I_d, \alpha, \left(\frac{n_2}{n_1}\right)_n, X_n\right)$$

in which U_d is the voltage on the direct current side.

The balance between the variables appearing in the three systems 1, 2 and 3 is obtained by an iterative method by successively solving these three systems.

c) In the models for studies on the dynamic behaviour of the power system, the representation of the converter stations is similar to that used for the alternating current distribution calculations. However, the transformation ratio of the transformers feeding the Graetz bridges shall be considered as fixed during the period of time covered by the simulation.

This is justified by the fact that action on the on-load tap changers is time-delayed to a relatively high value (of the order of ten seconds as a minimum).

For dynamic operating investigations, one can ignore the mode of action of the controls. The transfer functions and limitations on the specified power or constant current control loops are therefore represented.

The gains and time constants of the transfer functions of these controls are generally obtained by identification from the actual systems. The variations in the parameters: currents, voltages, and firing angles of the valves, are obtained by integrating the differential equations which govern the changes of these various parameters over a period of time.

FURTHER READING

Adam P., Collet Billon V. and Taisne J.R. (1986). La modelisation des liaisons à courant continu (The modelling of direct current links), *Revue Epure EDF*, September.

Arrillaga J. (1983). *High Voltage Direct Current Transmission*, Peter Peregrinus Ltd.

Cladé J. and Prijent M. (1970). Les principles de fonctionnement des transports à courant continu haute tension (The principles of operation of high voltage direct current transmissions). *Revue Générale d'Electricité*, May.

IEEE Working Group on 'Dynamic performance and modelling of dc systems', Hierarchical Structure.

Kauferle J., Koelsch H. and Sodek K. (1984). AC representation relevant to AC filtering and overvoltages for HVDC applications, CIGRE Report 14-08, Paris.

Liss G. (1989). A review of control functions in modern HVDC systems. CIGRE Colloquium on HVDC, Recife, Brazil, August.

Scott B. (1971). Load flow for ac and integrated ac–dc systems, PhD Thesis, University of Manchester.

Uhlmann E. (1975). *Power Transmission by Direct Current*, Springer Verlag.

Woodford D.A., Gole M.M. and Menzies R.W. (1983). Digital simulation of DC links and AC machines. *IEEE Trans*, **3**(3), 102.

Index

Page numbers appearing in **bold** refer to figures.

Adams method 256
Algebraic–differential equations
  5, 246, 247, 250
Algorithms 250
  digital models 5
  Electricité de France training
    simulator 217–20
  Gear's 256
  globally convergent 46–9
  integration 247–8
  recursive quadratic programming
    46–52
  superlinearly and locally
    convergent 49–52
Alternating current
  calculation tables 3
  transmission systems 243, 260
Alternators
  electromagnetic transients 138,
    140, 154, 167–74, 175
  lumped element representation
    153
  Park's equations 3, 167, 171, 175
  transient stability 100–11
  see also Synchronous machines
Ampere's theorem, transformers
  160
Analog models 3
  high-speed phenomena 3
  hybrid models 5
  Park's equations 3
  real time simulation 3, 208–9,
    225–6

replacement 3
Arc furnaces
  circuit diagram **184**
  random harmonic loads 183–5
Architecture, see Information
  technology architecture
Artificial intelligence 241
Asynchronous motors
  impedance 112–15
  load modelling 112–15
  short-circuit calculations 77
  single cage 112–15
  transient stability 112–15
Automatic control equipment
  test simulators 6
  training simulators 221
Automatic differentiation
  257–61
Axial power imbalance, two-zone
  neutron model 10

Behaviour models 7–10
  Lilliam model 8–10
  pressurized water reactors 8–10
Benders method, recursive
  quadratic programming 43–6
Bergeron's method
  electromagnetic transients
    174–5
  transmission line simulation
    226, 227–8
Bifurcation theory, nonlinear
  systems 251–3

INDEX 279

Black boxes, *see* Behaviour models
Blondel diagrams 110, 133
Boilers
   drum 88–91
   long term dynamics model 80–97
   power station 7–10
Busbars, harmonic propagation 191–8

Cables, *see* Lines and cables
Calcul de Reseaux Implicitement Couples (CRIC), *see* Coupling
Calculation tables
   reduced-scale models 3
   replacement 3
Capacitors, shunt 190–5
Circuit breakers 245
Coal preparation, thermal generating sets 91
Compensation elements
   negative resistances 2
   static VAR 115–17
   synchronous 117
   transient stability 115–17
Computing facilities 231–42
   Convex computers 229
   modeller–solver–display devices 253–5, 263–6
   multimedia 241
   networks 5, 233, 235–9
   servers 235–8
   workstations 233, 235–7, 240–1
Conductivity, lines and cables 141–2
Constrained optimization
   load flow calculations 33–46
   voltage profile 37–46
Consumption
   load modelling 111–15
   long term dynamics 79–97

steady state operation modelling 17
Control equipment, *see* Automatic control equipment
Coupling 28–30, 243–5
   CRIC decoupling technique 28–30
Current
   harmonic conditions 177–206
   short-circuit calculations 63–6
Cymharmo model, harmonic impedance 182

Damping circuits
   rotating machines 131–3
   transient stability 104, 114–15, 129–33
Data processing
   digital models 5
   information technology architecture 231–4
   object-oriented techniques 237
   technical data management 233–4, 249
Databases 237–8
Decision Tree Transient Stability method 251
Decomposition–coordination methods
   flow chart **45**
   recursive quadratic programming 43–6
Decoupling techniques, *see* Coupling
Differential equations 246–8
Digital computers, *see* Computing facilities
Digital models 3–4
   advantages 4–5
   algebraic–differential systems 5
   algorithms 5
   computer networks 5

data processing 5
probabilistic planning 5
Digital real time simulation
    208–30
  automatic controllers 221
  dispatching operations 209–24
  Electricité de France simulator
    215–24
  information technology
    architecture 229–30
  limitations 212–13
  parallel processing 227–30
  realism 213–14
  software 220–1
  switching devices 221
  test simulators 208–9, 224–30
  time measurement 210–11
  training simulators 208–24
  transmission lines 226–7
  validity 214
  see also Real time simulation
Direct current
  calculation tables 3
  high-voltage systems 243–5
  link modelling 270–7
  steady state operation 30–3
  transmission systems 30–3
  Zollenkopf method 32
Dispatching operations
  digital real time simulation
    209–24
  Electricité de France simulator
    215–24
Distribution systems
  harmonic propagation 190–8,
    205–6
  structure **190**
Domestic appliances, harmonic
    loads 183, 186–7
Drum boilers, long term dynamics
    model 88–91
Dynamic phenomena
  coupling 28–31, 243–5

parallel processing 238
see also Electromagnetic
  transients; Long term
  dynamics model; Transient
  stability

Eddy currents
  synchronous machines 167
  transformers 164–5
Electricité de France training
  simulator 215–24
  algorithms 217–20
  constraints 215–16
  topology 217, 221
  uses 221–3
  validity 216–18
Electromagnetic transients
  alternators 153
  Bergeron's method 174–5
  braking torque 138
  distributed element
    representation 138
  ferroresonance 139, 165–6,
    251–3
  high-speed phenomena 137–76
  impedance 137–8
  Kirchhoff's laws 140–1, 162,
    174
  lightning strikes 138–9
  lumped element representation
    138
  Maxwell's equations 140–1,
    160, 175
  nonlinear devices 174
  overhead lines and cables 138,
    140–59
  parallel processing 238
  rotor desynchronization 137–8
  subsynchronous resonance
    139–40
  switching operations 139
  synchronous machines 138, 140,
    167–74

INDEX   281

transformers   138, 140, 153, 160–6
tripped circuits   139, 245
see also Transient stability
Electromechanical oscillations
  stability   96–136
  time constants   1
Engineers
  data processing   231–4
  graphical user interfaces 239–41
Engines, transient stability   100–3
Equations
  modelling   7
  see also Park's equations; Power system equations; Telegraphic equations
Equipment rating, short-circuit currents   57–8
Equivalent models   4–5
Ethernet networks   236–8
EUROSTAG software   10, 248, 250, 255–66
Extended Equal Area Criterion method   251
Extended real time   211, 223–4

Fast transient conditions   1
  network analysers   2, 246
Faults
  asymmetrical   127–9
  clearing time   243–5
  symmetrical   127, 128–9
  transient stability   126–9
  see also Short-circuit currents
Feedwater plant, Lilliam model 9–10
Ferranti effect, lines and cables 155
Ferroresonance   251–3
  electromagnetic transients   139
  transformers   165–6
Flexible real time   211, 223–4

Freeware   238
Frequency
  lines and cables   138, 140–9, 156–9
  long term dynamics   79–97
  mean   137
Function models   5

Gauss–Seidel method, load flow calculations   25
Gear's algorithm   256
Generalized Minimal Residual method   250
Generating units
  asymmetrical faults   127–9
  damping circuits   104, 114–15, 129–33
  digital real time simulation 217–24
  harmonic conditions   183
  hydroelectric   94–7
  long term dynamics model 79–97, 83
  pressurized water reactors   84, 91–4
  steady state operation   17
  thermal generating sets   89–91
  transient stability   100–11
  voltage regulators   107–11
Graetz bridges   270–3
Graphical user interfaces   231, 237, 239–41
  artificial intelligence   241
  historical development   240–1

Harmonics   177–207
  busbars   191–8
  Cymharmo model   182
  distribution systems   190–8, 205–6
  generators   183
  harmonic disturbance   177, 188
  impedance   178–83, 187–205

lines and cables   204–5
loads   183–7, 193–5, 197–8, 206
propagation   188–202
shunt capacitors   190–5
source modelling   177, 183–8
substations   190–8, 205–6
telegraphic equations   199–200
Thevenin diagram   **179**
transformers   183, 191, 202–5
transmission systems   198–202
High-speed phenomena
  analog models   3
  electromagnetic transients
    137–76
  hybrid models   5
  real time simulation   3, 217–24
High-voltage systems
  direct current   243–4
  short-circuit clearing time   243–5
Hybrid models   4–5
Hydroelectric power units
  long term dynamics model
    94–7
  water constant   96–7
Hypertext   241

Impedance
  asynchronous motors   112–15
  Cymharmo model   182
  electromagnetic transients
    137–8
  harmonic conditions   178–83,
    187–205
  internal   149–56
  lines and cables   149–56
  measurement   178–81
  Thevenin diagram   **179**
Information technology
  architecture   231–9
  data processing   231–4
  digital real time simulation
    229–30

distributed   239
electrical systems   233–4
freeware   238
hardware   237
modular design   237
networks   5, 233, 235–9
office systems   241
parallel processing   238–9
Power System Information
  Systems   233–4
servers   235–8
software   237–9
workstations   233, 235–7, 240–1
World Wide Web   238
Integration
  method choice   255–7
  variable stepsize techniques
    247–8
Islanding, pressurized water
  reactors   10
Iterative methods   250

Kirchhoff's laws
  direct current links   276
  electromagnetic transients
    140–1, 162, 174
  lines and cables   140–1
  transformers   162
Knowledge models   7–8

Lenz's law
  synchronous machines   167–8
  transformers   160
Lightning, electromagnetic
  transients   138–9
Lilliam model
  EUROSTAG software   10
  general representation   **9**
  pressurized water reactors   8–10
  protection activation   8
Lines and cables
  Bergeron's method   226, 227–8

conductivity 141–2
digital real time simulation
  226–7
distributed element
  representation 138, 140–59
electromagnetic transients 138,
  140–59
Ferranti effect 156
frequency range 142, 143–9,
  156–9
geometry 141–2
harmonic conditions 204–5
internal impedance 149–54
Kirchhoff's laws 140–1
long term dynamics model 83
Maxwell's equations 140–1
natural modes 154–6
overhead 66, 138, 140–60
overvoltages 156
steady state operation modelling
  13–14
telegram operators equations
  149, 154–6, 226
time domains 156–60
transient stability 118
Load flow calculations
  constrained optimization 33–46
  CRIC decoupling technique
    28–30
  Gauss–Seidel method 25
  long term dynamics 79–97
  Newton method 25–8, 29–30
  nonlinear systems 21–28
  short circuits 77–8
  Sparse–Broyden method 28
  steady state operation 20–30,
    33–46
  transformers 20–31, 66–7
  transmission systems 20–33
Loads
  arc furnaces 183–5
  asynchronous motors 112–15

harmonic conditions 183–7,
  193–5, 197–8, 206
random 183–5
semiconductor-based devices
  183, 185–7
transient stability 111–15
Long term dynamics model 79–97
  boilers 80–97
  calculation methods 80–3
  component modelling 83–9
  coupling 245–7
  EUROSTAG simulation 261–2
  generating units 83–4, 89–97
  loads 84–6, 89
  Newton–Raphson method 82–3
  shaft lines 80–97
  stability 247–8
  training simulators 83, 217
  turbines 80–97

Magnetic circuit saturation, *see*
  Saturation
Magnetization, transformers
  161–5
Maxwell's equations
  electromagnetic transients
    140–1, 160, 175
  lines and cables 140–1
  transformers 160
Mean frequency 137
Modeller–solver–display devices
  253–5, 263–6
Modelling 6–11
  behaviour models 7–10
  constraints 6
  data quality 11
  harmonic conditions 177, 183–8
  knowledge models 7–8
  lines and cables 13–14
  nodal topology 13
  power station boilers 7–10
  short-circuit currents 66–78

284 INDEX

steady state operation 12–17
validity 6
see also Real time simulation
Motif software 237, 241
MS-DOS software 241

Networks
 symmetrical 60–4
 see also Computing facilities
Newton's method
 load flow calculations 26–7, 29–30
 transformers 164
Newton–Raphson method 82–3, 247
Nonlinear systems
 bifurcation theory 251–3
 electromagnetic transients 174
 load flow calculations 21–8
Nordsieck's formulism 247, 255–6
Nuclear power units, pressurized water reactors 8–10, 84, 91–4

Operation security, reactive power compensation devices 39–46
Optimization
 constrained 33–46
 recursive quadratic programming 36–52
Oscillations
 nonlinear 251–3
 see also, Electromechanical oscillations
Overhead lines and cables
 electromagnetic transients 138, 140–59
 short-circuit calculations 66
Overvoltages, lines and cables 155

Parallel processing
 digital real time simulation 227–30
 dynamic phenomena 238
 electromagnetic transients 238
 information technology architecture 238–9
 short-circuit currents 238
Park's equations
 alternators 3, 167, 171, 175
 synchronous machines 104, 109, 167, 171, 245
PC workstations 237
Power converters, harmonic loads 184, 185–6
Power systems
 equations 17–19
 information systems 233–4
Pressurized water reactors
 900 MW 8–10, 84, 91–4
 islanding 10
 Lilliam model 8–10
 long term dynamics model 91–4
 torque 8–10
 two-zone neutron model 10
Probabilistic planning 5
Protection equipment
 circuit breakers 245
 test simulators 6
PWR, *see* Pressurized water reactors

Rating, *see* Equipment rating
Reactance, transformers 66–76
Reactive power compensation devices
 recursive quadratic programming 37–46
 security rules 39–46
Reactor coolant pumps, Lilliam model 8–10
Real time simulation
 analog models 3, 208–9, 225–6

digital models 208–30
high-speed phenomena 3
hybrid models 5
modelling constraints 6
Power System Information
  Systems 234
reduced-scale models 3
simulation time 249
see also Digital real time
  simulation
Recursive quadratic programming
  algorithms 46–52
  Benders-type method 43–6
  constrained optimization 36–52
  decomposition–coordination
    43–6
  reactive power compensation
    devices 37–46
  voltage profile optimization
    37–46
Reduced-scale models 2–5
Resistivity, *see* Conductivity
Resonance
  subsynchronous 139–40
  see also Ferroresonance
Rotating machines
  desynchronization 137–8
  transient stability 131–3

Saturation
  synchronous machines 125–6
  transformers 163–4
Security rules, *see* Operation
  security
Semiconductor devices, harmonic
  loads 183, 185–7
Shaft lines
  long term dynamics model
    80–97
  synchronous machines 173–4
Short-circuit currents 57–78
  asynchronous machines 77

busbars 191–8
calculation methods 60–6
clearing time 243–5
definition 58–9
electrical variables calculation
  63–4
Electricité de France training
  simulator 217–24
equipment rating 57–8
extremely high voltage 243–5
loads 77–8
network element modelling 66–
  78
overhead lines and cables 66
parallel processing 238
substations 190–1
symmetrical components 60–4,
  127, 128–9
synchronous machines 76–7, 171
Thevenin's theorem 60–4
transformers 66–76
Shunt capacitors, harmonic
  propagation 190–5
Single-phase circuits, steady state
  operation 12–56
Slow transient conditions, long
  term dynamics 80–97, 248
Small signal stability
  eigenvalues method 121–3, 125
  electromechanical oscillations
    98–100, 119–25
  transient stability model 119
  transmittance method 120–4
Software
  digital real time simulation
    220–1
  EUROSTAG 10, 248, 250,
    255–66
  information technology
    architecture 237–9
Sparse–Broyden method, load flow
  calculations 27

286 INDEX

Stability
  coupling   245–7
  differential equations   246–7
  electromechanical oscillations
    96–136
  monotonic instability   99
  oscillatory instability   99
  transmission systems   248–9
  see also Small signal stability;
    Transient stability
Star transformers, short-circuit
  calculations   67–76
Static converters, harmonic loads
  186
Static VAR compensators,
  transient stability   115–17
Steady state operation
  constrained optimization   33–46
  consumption   17
  direct current approximation
    31–3
  generation   17
  lines and cables   13–14
  load flow calculations   20–30,
    33–46
  power system equations   17–19
  recursive quadratic programming
    algorithms   46–52
  single-phase circuits   12–56
  system modelling   12–17
  time constants   1
  transformers   14–17
Steam generators
  pressurized water reactors
    8–10, 91–4
  water level   **10**
Steam turbines, transient stability
  101–3
Stokes's formula, transformers
  160
Substations
  harmonic propagation   190–8,
    205–6
  short-circuit currents   190–1
  topology   13
Subsynchronous resonance
  139–40
Subsystems, equivalent models
  4–5
Switching
  digital real time simulation
    221
  electromagnetic transients
    139
Symmetrical components, short-
  circuit current calculations
  60–4, 127, 128–9
Synchronous machines
  compensators   117
  eddy currents   167
  electromagnetic transients   138,
    140, 167–74
  Lenz's law   167–8
  lumped element representation
    138, 140, 167–74
  negative system   76–7
  nonlinearities   138, 140
  Park's transform   104, 109, 167,
    171, 245
  positive system   76
  saturation   125–6, 169
  shaft lines   173–4
  short-circuit calculations   76–7,
    171
  torque   171
  transient stability   103–11
  zero phase-sequence system   77
  see also Alternators

Tap changing, transformers   15–
  31, 84–7, 89
Technical data management, see
  Data processing; Power
  systems, information systems

Telegram operators equations
  lines and cables   149, 154–6, 226
  transmission systems   199–200
Television sets, harmonic loads   186–7
Test simulators
  modelling constraints   6
  real time   208–9, 224–30
Thermal generating sets
  coal preparation   91
  long term dynamics model   89–91
Thevenin's theorem
  harmonic systems   **179**
  short-circuit current calculations   60–4
Time
  constants   1, 2, 6, 243–7
  domains   156–9
  measurement   210–11
  response times   249–55
  scales   1, 243–7
  see also Digital real time simulation; Real time simulation
Topology
  Electricité de France training simulator   217, 221
  nodal   13
  substations   13
Torque
  electromagnetic transients   138
  pressurized water reactors   8–10
  synchronous machines   170–1
Training simulators
  dispatching operators   209–24
  Electricité de France   215–24
  long term dynamics model   83, 217
  modelling constraints   6

  real time   208–24
Transformers
  Ampere's theorem   160
  asymmetrical faults   127–9
  eddy currents   165
  electromagnetic transients   138, 140, 153, 160–6
  ferroresonance   165–6
  harmonic conditions   183, 191, 202–5
  internal resonance   160
  Kirchhoff's laws   162
  Lenz's law   161
  load flow calculations   20–30, 66–7, 84–7, 89
  long term dynamics model   83, 84–7, 89
  lumped element representation   138, 140, 153, 160–7
  magnetization   161–5
  Maxwell's equations   160
  Newton's method   164
  nonlinearities   138, 140
  power system equations   17–19
  reactance   66–76
  saturation   162–4
  short-circuit calculations   66–76
  steady state operation modelling   14–17
  tap changing   15–31, 84–7, 89
  three-phase winding   68–76, 160, 165–6
  transient stability   118
Transient conditions
  electromechanical oscillations   1
  fast   1
  slow   80–97, 248
  time constants   1
Transient stability
  alternators   100–11
  Blondel diagram   110, 133

compensation elements 115–17
coupling 245–7
digital real time simulation 220
direct simulation methods 250–3
Electricité de France training simulator 222–3
electromechanical oscillations 96–119
engines 100–3
EUROSTAG simulation 261–2
faults 126–9
generators 103–11
lines and cables 118
loads 111–15
long term dynamics 247–8
magnetic circuit saturation 125–6
network analysers 2, 246
power systems 118
rotating machines 131–3
small signal stability 119
solution methods 118–19
transformers 118
see also Electromagnetic transients
Transmission lines, see Lines and cables
Transmission systems
French **200**, 201
harmonic propagation 198–202
load flow calculations 20–33
stability 248–9
Tripping, electromagnetic transients 139, 245

Turbines, long term dynamics model 80–97
Two-zone neutron model 10

UNIX workstations 237, 241

Variable integration interval techniques 247–8
Voltage
constrained optimization 37–46
Electricité de France training simulator 222
harmonic conditions 177–206
long term dynamics 79–97
regulators 107–11
short-circuit current calculations 63–6

Water constant, hydroelectric power units 96–7
Windings
time constants 2
transformers 68–76, 160, 165–6
Windows software 237, 241
Workstations
information technology architecture 233, 235–7, 240–1
PCs 237
UNIX 237, 241
World Wide Web 238

X-windows software 237, 241

Zollenkopf method, direct current approximation 32